Y0-BLC-303

Metropolitan College of NY
Library - 7th Floor
60 West Street
New York, NY 10006

International Disaster Management Ethics

International Disaster Management Ethics

LIZA IRENI SABAN

SUNY PRESS

Metropolitan College of NY
Library - 7th Floor
60 West Street
New York, NY 10006

Published by State University of New York Press, Albany

© 2016 State University of New York

All rights reserved

Printed in the United States of America

No part of this book may be used or reproduced in any manner whatsoever without written permission. No part of this book may be stored in a retrieval system or transmitted in any form or by any means including electronic, electrostatic, magnetic tape, mechanical, photocopying, recording, or otherwise without the prior permission in writing of the publisher.

For information, contact State University of New York Press, Albany, NY
www.sunypress.edu

Production, Jenn Bennett
Marketing, Michael Campochiaro

Library of Congress Cataloging-in-Publication Data

Names: Ireni Saban, Liza, author.
Title: International disaster management ethics / Liza Ireni Saban.
Description: Albany : State University of New York Press, 2016. | Includes bibliographical references and index.
Identifiers: LCCN 2015042427 (print) | LCCN 2016011153 (ebook) | ISBN 9781438461717 (hardcover : alk. paper) | ISBN 9781438461724 (e-book)
Subjects: LCSH: Disaster relief—International cooperation. | Emergency management—International cooperation. | Disaster relief—Moral and ethical aspects.
Classification: LCC HV553 .I5887 2016 (print) | LCC HV553 (ebook) | DDC 174/.936334—dc23
LC record available at http://lccn.loc.gov/2015042427

10 9 8 7 6 5 4 3 2 1

Contents

Introduction		1
Chapter 1	Introduction to International Disaster Management Ethics	11
Chapter 2	Dilemmas and Ethical Issues in International Disaster Management	25
Chapter 3	Global Distributive Justice	39
Chapter 4	The Dependency Syndrome	63
Chapter 5	Donation Fatigue	77
Chapter 6	Corruption	89
Chapter 7	Compensation	103
Chapter 8	Code of Ethics for the International Disaster Management Practice	121
Conclusion		145
Notes		151
References		157
Index		179

Introduction

With the increasing incidence and intensity of natural disasters, the international disaster management community has its hands full. Its time-constrained environment is complex and complicated, involving multiple responsibilities, challenges, and constituencies. International aid actors' ability and commitment to meet immediate human need in times of disaster is tested during disaster events as they try to improve on the affected national governments' ability to address the needs of the affected population by supplying relief resources and services. Large-scale disasters, often termed international disasters, overwhelm national governments' capacities to respond. This follows the UN definition of natural disaster as[1]: "The consequences of events triggered by natural hazards that overwhelm local response capacity and seriously affect the social and economic development of a region."[2]

The frequency and severity of natural disasters is increasing. During 2000–2014, 6,084 natural disaster events were registered. The annual average of disaster frequency observed is about 400 natural disasters per year. Natural disasters caused the death of 1,218,296 people, affecting 2,919,832,577 people worldwide.[3] Total reported economic damage from natural disasters is estimated at US$ 167 billion.[4] Natural disasters affect both developed and developing countries across the globe; China, the United States, the Philippines, India, and Indonesia are considered the five countries that are most frequently hit by natural disasters.

In these extreme events, response and relief efforts are quickly mobilized by the international disaster management community. The international disaster management community involves a wide range of actors embedded in fluid and complex networks of association. The international disaster management community consists of victims, governments of the affected countries, governments of other countries and their emergency agencies, international organizations, regional organizations, and NGOs.

Too often, national governments accuse international relief actors of overlooking the local context in which they operate, and even that their underlying goal is to broaden their agendas and question the priorities and methods of the elite-controlled political processes (Aldrich and Crook 2008). Although international relief organizations may not have a full understanding of the wider socio-economic context in which response and relief actions occur, evidence from various disaster settings points to lack of confidence on the part of international aid organizations regarding the capacity of the affected government to effectively deliver relief aid to its citizens. This mutual distrust turns into greater inconsistency in the disaster policy implementation process, thus failing to manifest effective disaster response (Adams and Balfour 2009, Ink 2006, Jia 2008, Menzel 2006, Mooney 2009, Polman 2010, Yang 2008).

Nonetheless, there is another set of challenges faced by the international disaster management community that is equally implicated in effective disaster management: The potential for disunity, when considering the vast differences in international aid actors' values, preferences, and institutional settings.

The international disaster management community is composed of innumerable agencies with different skills, reporting channels, level of responsibility, and ethical cultures. In a disaster management setting, international aid organizations must function in the midst of competing claims about the distribution of resources, conflicting views of the management of personnel, scarcity of appropriate information, and lack of coordination. It seems that Kevin M. Cahill's statement that "Managing complex humanitarian emergencies, particularly in the midst of conflicts and disasters, is not a field for amateurs" (Cahill 2013, 385) is truer than ever.

Viewed in this way, universalizing professional and ethical obligations through a process of codification may help to build trust with the disaster-affected government and tackle the unintended consequences of immediate short-term international response for people affected by natural disasters. Studies have shown that in the face of a disaster event, effective coordination compensates for scarce resources and uncertainty (Comfort and Haase 2006, Comfort, Ko, and Zagorecki 2006, Coppola 2011, Drabek 2003, Kapucu 2006, 2008, Kapucu, Arslan, and Collins 2010, Kapucu, Augustin, and Garayev 2009, Kobila, Meek, and Zia 2010, McEntire 2002, Mitchell 2006, Moynihan 2012, Nolte, Martin, and Boenigk 2012, Vasavada 2013). Professionalism through the process of codification in international disaster management then acts to establish the rules, norms, and assumptions that shape the distribution of aid resources by the overall international aid community. Professionalism provides controls that regulate how international aid actors manage ambiguity and ethical dilemmas.

A strategy for universalizing professional understandings of ethics in international disaster management is designed to clarify and identify international disaster management professions across a range of organizational and national contexts. Clearly, consideration of ethics in making international disaster management professional must ensure that this institution is just. During and after a disaster event, the international disaster management community is faced with ethical dilemmas regarding appropriate and fair allocation of relief and response resources and services. While it is desirable for the management regime to promote immediate and efficient distribution of response and relief resources, long-term distributional consequences often go unrecognized within the overwhelming, time-constrained environment created by disaster events. The fact that the international disaster management community intervenes in distributional decision making with varying degrees of legitimacy and political will, leads to having special responsibilities to maintain the highest ethical standards in its role as enabler, coordinator, and facilitator to meet human need in times of disaster. It is suggested then that humanitarian aid and relief organizations are not morally neutral agents judged on their effectiveness or reduction of disaster vulnerabilities, but rather moral agents that serve cosmopolitan values, and are concerned with justification of the ends and means by which those values are enacted. This is the basis of the international disaster management community's claim to moral legitimacy in governance.

It is argued that the recasting of ethical responsibility in terms of unintended and indirect consequences of distribution of relief resources constructs ethics of international disaster management on global interconnectivity. More troubling than the need to address immediate needs and quickly deliver humanitarian aid and relief, however, are the long-term implications of the international disaster management community's emergency efforts. Thus, large-scale disasters make international aid organizations more reliant on coordination and thus reliant on the establishment of common principles, common goals, and mutual actions. For that, a uniform set of ethical guidelines for international aid actors must articulate the responsibility to meet the demand of global distributive justice.

In making visible professional responsibility for international disaster management practitioners, global distributive justice has the capacity provide common ground to raise consciousness about the demanding political and ethical challenges faced by the international disaster management community. This book draws on current prominent perspectives on global justice to international disaster management, such as the consequence-oriented approach developed by Peter Singer, the rights- and institutions-oriented approach of Thomas Pogge, and the capabilities-oriented approach of Amartya Sen, drawing examples from recent large scale natural disasters, namely the 2004 Indian Ocean earthquake and tsunami, the 2005 Gulf Coast Hurricanes (United States), the 2010 Chile earthquake, the 2010 Haiti earthquake, and the Philippines typhoon (2013). The book concludes with a Code of Ethics that covers the ethics spectrum, touching on the major areas of concern in international disaster management to combine management and leadership for international disaster management ethics.

The role humanitarian aid and relief organizations could play in creating a cosmopolitan condition in international disaster management lies in the blurred lines between management and leadership. The present book offers to transcend the distinctions between management and leadership commonly entrenched in traditional disaster management literature. In response to this challenge, this book offers to create stronger links between cosmopolitan theory and contemporary international relief practice.

By doing so, this book maps out the recent relationship between well-established theories of global justice and institutional arrangements set by humanitarian aid and relief organizations engaged in international disaster management such as the Code of Conduct for the International Red Cross and Red Crescent Movement and NGOs in Disaster Response Programmes, the Humanitarian Charter and "The Sphere Project," the Humanitarian Accountability Partnership (HAP), "People in Aid Code of Good Practice," and the "Good Humanitarian Donorship" initiative. Codes of ethics are considered an important management tool for promoting international disaster management to meet global justice criteria. The present book urges that studies of international disaster management be explicitly and systematically grounded in ethical considerations embedded in a global justice framework that could set a precedent for future decision making by relief organizations and national government agencies. The relevance of contemporary theories of global justice to international disaster management practices defines the special moral role of humanitarian aid and relief organizations in disaster emergency management as part of the extended duty of assistance.

The originality of this book rests in enhancing international disaster management capacity to professionalize distributive decision making of its diverse members through the process of codification. Rather than promoting ethical international disaster management as an actor's obligation, taking into account global distributive considerations becomes part of defining the profession at the global level in the face of the growing number and size of natural disasters' effects. The book suggests that it is crucial for international aid organizations engaged in disaster management to attempt to lift the moral fog that envelops their practice and to alert to the ethical implications and meaning of their decisions and actions, commitment to exercise ethical judgment, and leadership. International disaster management needs to claim control over its professional evolution and take the lead in communicating members' expectations regarding the standards of relief provision. Finally, the book complements a continued focus on disaster management practices by reinvigorating the ethical aspects and issues surrounding disaster management that are greatly debated at the global level.

The Book's Outline

In chapter 1 we provide a valuable and thoughtful understanding of ethical issues related to international disaster management that require important ethical decision making. The chapter shows how the growing incidence of natural disasters throughout the world has brought new challenges to the international disaster management community. These challenges cannot be resolved unless the international disaster management community recognizes the importance of ethics as part of their profession, just as any other aspect such as efficiency, service quality, costs (budgeted and actual), personnel evaluation, and timeliness. Dealing with such performance indicators is to be on safe ground. If we move to deal with ethical dilemmas and challenges, we pass through an unsettling or even risky territory. This view is articulated by the spokesperson for the International Committee of the Red Cross (ICRC) who discussed the boundaries of the humanitarian work terrain detached from broader issues of social justice: "We do surgery. We do medicine. We do clean water. We don't do justice" (Nickerson 1997). Or as stated by James Orbinski while receiving the Nobel Peace Prize on behalf of Doctors Without Borders back in 1999: "Humanitarianism is not a tool to end war or to create peace. . . . It is an immediate, short-term act that cannot erase the long term necessity of political responsibility" (Orbinski 1999).

Against these views, it is suggested that the ethical facet is intensified in disaster events when international aid agencies become bound to their own standards of conduct since they operate outside the areas of established public law. Thus, chapter 1 presents the relationship between international disaster management as an institutional structure and social justice. By drawing on existing institutional aspects of the international disaster management system such as rules, norms, best practices, professional guidelines, and Codes of Ethics/Conduct, it is evident that although international humanitarian aid organizations share the responsibility to reduce suffering and provide immediate response for people affected by disasters, there is a potential for disunity when considering the vast differences in actors' institutional structures, standards, and operational strategies.

Chapter 2 outlines some ethical dilemmas and issues related to international disaster management. A moral or ethical dilemma refers to a situation in which an agent is faced with two conflicting moral obligations where neither moral obligation overrides the other. Thus, the conflict is irresolvable. Moral or ethical dilemmas are amplified under conditions of disaster in which multiple, pressing needs increase and time constraints force immediate choices within a limited time frame to assess the full range of potential consequences of each decision or action. Institutional challenges may also challenge the delivery of humanitarian aid in disasters. International aid organizations may be unable to effectively respond to numerous and pressing needs due to multiple administrative and policy barriers. Thus, this chapter briefly introduces four kinds of ethical dilemmas frequently faced by the international disaster management community. We do not provide an exhaustive list of the ethical dilemmas encountered by the international disaster management community, but rather a few ethical dilemmas arising regarding the allocation of response and relief resources that haunt the global community including the dependency syndrome, donation fatigue, corruption, and climate change compensation.

Chapter 3 attempts to bridge the previous chapter and the subsequent chapters by developing a comprehensive framework for ethical decision making in international disaster management. In this chapter we suggest that in times of disaster the duty of assistance may be limited to immediate relief based on the institutional capacity for allocative decision making made by international aid organizations engaged in response and relief efforts. In sudden-impact situations, it is likely that the long-term consequences of the allocative decisions and actions made by humanitarian aid organizations may go unnoticed. Viewed in this way, distribution of disaster resources must assert the primacy of humanity over the sovereignty of states, thus leading to holding principles of cosmopolitan justice applicable beyond national borders. Over the years, proponents of cosmopolitan justice have attempted to develop an acceptable framework for validating the duty to assist other peoples living under unfavorable conditions and a globally shared responsibility. This chapter introduces three well-established theories of global justice including the consequence-oriented approach developed by Peter Singer, the rights- and institutions-oriented approach developed by Thomas Pogge, and the capabilities-oriented approach of Amartya Sen. The ethical analysis presented in this chapter yields cosmopolitan principles that will be applied to contemporary ethical dilemmas faced by the international disaster management community discussed in the following chapters.

Chapter 4 addresses the dilemma of the dependency syndrome. The mandate of humanitarian aid and relief organizations should be to supplement affected national governments and their available resources. However, dependency syndrome often results from the reliance of the disaster-affected population (receivers) on humanitarian aid and on its providers. The idea of dependency syndrome in disaster emergency contexts refers to the potential negative impacts of humanitarian aid or emergency relief. Although international relief aid used in humanitarian emergencies aims at reducing vulnerabilities, it may lead to unintended, adverse consequences associated with weakening local capacity and the responsibility of government agencies to deliver aid services and resources to meet basic humanitarian needs. This chapter brings some clarity about the conceptualization of dependency syndrome when used within the field of international disaster management. We further argue that the acceptance of the negative aspects associated with the notion of dependency may justify a withdrawal from response and relief efforts, or on governments to claim overall control of the whole response and relief process and means. Thus, the chapter develops the notion of dependency syndrome as referred to by global theories of justice to be usefully applied to provide justifications for action or inaction in the context of international disaster management.

Chapter 5 discusses the dilemma of donation fatigue. Donor fatigue is caused by budget exhaustion. In an era of tightening budgets, and the growing number and size of natural disaster events, donation funding raised by humanitarian aid and relief organizations has started to fall. When multiple large-scale disaster events occur concurrently, or in quick succession, humanitarian aid organizations face difficulty in concentrating on more than one humanitarian aid effort at a time. The disasters of

2004–2005, for example, demonstrate that timing of a disaster event in relation to other events can profoundly impact the response assistance provided by the international community.

On December 26, 2004, a massive earthquake struck the west coast of northern Sumatra, Indonesia, followed by severe aftershocks. The international response to the crisis was significantly high, as over US$7 billion was raised from international donations. (Telford and Cosgrave 2006) On March 28, 2005, the west coast of Sumatra and the islands of Nias, Simeulue, and Banyak faced a powerful aftershock that caused massive damage. (Telford and Cosgrave 2006) In July 2004, heavy monsoon rains led to extensive flooding in a large area of Bangladesh. The floods killed more than 600 people, left nearly 1.7 million people displaced, and resulted in massive damage affecting the lives of 36 million people. The Bangladesh aid appeal fell short of its target as international donor contributions to Bangladesh were only $28.2 million (Bangladesh DER 2004). The Caribbean hurricanes, namely Charley, Frances, Ivan, and Jeanne, led to 3,258 fatalities between August and October 2004 (Zapata Martí 2005).

In November–December 2004 a series of typhoons and storms were reported in the Philippines, leading to over 1,000 deaths and causing massive damage on the island of Luzon.[5] In January 2005, coastal regions of Guyana, including the capital, Georgetown, faced devastating flooding affecting over 300,000 people.[6] Early in 2005 most of the Central Asian region including Afghanistan, Pakistan, and Tajikistan suffered from heavy rains. In Pakistan alone over 600,000 people were affected and 486 people died due to the flooding. The international community responded with more than US$4 million of assistance to Pakistan (UNDP, 2005). On February 22, 2005, an earthquake struck the Zarand district in the Kerman province of the Islamic Republic of Iran. The earthquake caused the death of 612 people, with massive damage affecting more than 30,000 people. Although no appeal was made by the Government of the Islamic Republic of Iran, several United Nations agencies, such as UNICEF and WHO, delivered immediate relief assistance.[7] In August 2005, Hurricane Katrina struck the Gulf Coast; international donors provided over $1 billion of financial assistance and relief services.[8]

Other natural disasters, health epidemics, drought, and famine were evident in developing countries during this period; however, they did not generate international headlines such as the Indian Ocean tsunami that attracted great international attention and generous response aid. In a similar vein, in the 2010 earthquakes in Haiti, Chile, and Taiwan, the response to both the Chile and Taiwan crises was significantly less than to

Haiti just weeks prior (Daniell 2011). These examples demonstrated that "low-profile" disasters such as drought, erratic rains, locust infestation, and floods suffer from lesser international attention and commitment despite the fact that they have extensive consequences on development in poor and least developed countries. In other instances, humanitarian aid and relief organizations may grow frustrated with constant appeals for donations. Frustration may also be intensified when the international community donates to national governments that mismanage loans and donations, resulting in political and administrative corruption and underdevelopment coupled with high indebtedness.

Chapter 6 engages with the dilemma of the allocation of international relief resources and corruption, especially in poor countries affected by disaster. Humanitarian aid and assistance involve large amounts of cash and supplies that may create opportunities for corruption especially in poor countries. Relief is delivered in challenging environments. The introduction of large amounts of resources into poor developed countries during disaster response and relief phases may amplify power imbalances and create more opportunities for corruption. In addition, existing levels of perceived corruption in countries may halt international aid organizations' motivation to mobilize disaster response and relief efforts to these countries due to the risks of aid being diverted by elite groups and embedded corrupt systems. Thus, this chapter discusses the way corruption in the allocation of aid undermines the moral obligation behind humanitarian assistance, leading to inequitable and ineffective distribution of disaster relief and reconstruction aid.

Chapter 7 addresses the claims of poor nations for compensation for disaster-related losses from pollution by developed countries. The recent devastating typhoon that caused the death of thousands of people in the Philippines has fueled the debate about whether rich nations should compensate poor ones for climate-related losses. The inevitable results of global warming and climate change are among many natural disasters, e.g., cyclones, hurricanes, typhoons, floods, volcanic eruptions, earthquakes, and tsunamis, which nations all over the world are experiencing. Since developed and industrialized nations are considered the greatest producers of greenhouse gases (GHG), they should be held liable for climate change. However, an obvious difficulty in justifying such liability in terms of global distributive justice is that given the uncertainty about the exact causes and consequences of global warming, it will be very difficult to assess that natural disasters result from GHG emitters. On the other hand, the industrialized and developed nations can equally accuse the less developed nations of causing environmental damages from destruction of forest areas. Does this evidence still justify these nations' claims for compensation, too?

Chapter 8 explores the way to improve ethical decision making and conduct by codification of professional ethics in international disaster management. This means that all international response and relief organizations will use the same professional ethics, even though each organization is guided by its own practical standards, which vary somewhat from organization to organization. When faced with an ethical dilemma associated with disaster management, it is important to remember that there is seldom only one right way in which to act. However, this chapter intends to sets forth a Code of Ethics that covers the ethics spectrum, touching on the major areas of concern in international disaster management as a guide to assist international disaster management practitioners to make professional and ethically responsible decisions that meet global justice criteria.

This chapter introduces a specific codification of those ethical principles underpinning the global distributive justice theories set by Peter Singer, Thomas Pogge, and Amartya Sen that have the potential of producing a more equitable distribution and access to response and relief resources. It is intended to serve as a unified framework for good practice in international disaster management, and to inspire the will of members of the international disaster management community to act in a manner consistent with those tenets. The international disaster management community is expected to take into consideration all principles in this Code that have a bearing on any situation in which ethical judgment is to be exercised, and to select a course of action consistent with the global justice criteria set by the Code.

1

Introduction to International Disaster Management Ethics

> International disaster management has become increasingly diverse, encompassing new areas of technical expertise not traditionally considered relevant to the profession.
>
> —Coppola 2011, 641

The growing incidence of natural disasters throughout the world has brought new challenges to the international disaster management community. The present chapter reviews concerted efforts to create shared institutional frameworks to form a basis of collective standards and behaviors in delivering international aid in emergencies. The aim of the chapter is to consider the extent to which the international disaster management community has a moral responsibility to address the broader implications of its immediate allocative decisions and actions in the face of adversity. It is suggested that international aid allocation is only one aspect of the interface between ethics and politics in international disaster practice.

International Disaster Management Regime

International disaster management often refers to designating the efforts of a global community of responders to assist the affected nation or nations in their disaster response efforts. The scale of the disaster dictates the range of response and recovery needs (Coppola, 2011). Extreme events overwhelm national governments' capacities to respond, and force governments of the affected nations to call upon the resources and services of the international disaster management community outside their hierarchical control. In these cases, response efforts are centered on the international disaster coordination system to quickly mobilize response resources and assist affected populations to effectively manage disaster relief and risk reduction in such a short time frame (Comfort and Haase 2006, Comfort, Ko, and Zagorecki 2006, Drabek 2003, Kapucu 2006, 2008, Kapucu, Arslan, and Collins 2010, Kapucu, Augustin, and Garayev 2009, Kobila, Meek, and Zia 2010, McEntire 2002, Mitchell 2006, Moynihan 2012, Nolte, Martin, and Boenigk 2012, Vasavada 2013).

Since 1990, natural disasters have affected about 217 million people every year (Guha-Sapir, Vos, and Below 2012). Natural disasters result from various causes including geophysical (earthquakes, landslides, tsunamis, and volcanic activity), hydrological (avalanches and floods), climatological (extreme temperatures, drought, and wildfires), meteorological (cyclones and storms/wave surges), and biological (disease epidemics and insect/animal plagues).[1] Based on a forecasting model created by Oxfam, by 2015 over 375 million people on average per year are likely to be affected by climate-related disasters.[2] This number exceeds 50 percent more than have been affected in an average year during the last decade. Increased occurrence and intensity of natural disasters during the last decade have significant impact on people directly and indirectly including death, disabilities, and disease outbreaks.[3] Direct impacts of natural disasters refer to mortality and injury, damage to infrastructure, damage to homes and contents, damage to firms, and environmental degradation, while indirect impacts include costly adaptation or utility reduction from loss of use, mortality, morbidity, and business interruption (Rose 2004).

For example, the earthquake in Haiti in 2010 and Cyclone Nargis in Myanmar in 2008 caused the death of 225,000 and 80,000 people, respectively, and immense numbers of injuries, illness, and property damage. Although estimating the full range of economic costs from natural disasters is difficult, the damages from natural disasters have risen from an estimated $20 billion on average per year in the 1990s to about $100 billion per year during 2000–10.[4] According to a recent IMF study, this trend is expected to grow due to the rising concentrations of people living in areas most vulnerable to natural disasters and climate change (Laframboise and Loko 2012, 1–31). The Great East Japan earthquake, which occurred in March 2011, caused Japan an estimated direct economic cost of 16.9 trillion yen ($210 billion), which is also calculated at 3.6 percent of 2011 GDP. The earthquake has led to immense destruction of roads, railways, airports, schools, and other infrastructures (IMF, 2012).

Evidence suggests that there are some communities that are more prone to hazards. Since the 1960s, an estimated 99 percent of the world's population has been affected by disasters and 97 percent of all fatalities have occurred in middle- and low-income countries (Laframboise and Loko 2012). In addition, disasters lead to annual economic losses in developing countries that amount to nearly 2 to 15 percent of their GDP (United Nations 2005, 181). The trend of rapid urbanization, for example, has led to poorer people being marginalized from safe and legal areas in many developing countries, which leaves communities at high disaster risk. The combination of increased number and severity of natural disasters with diminished coping mechanisms of an affected population raises the need for reliance on international disaster response and relief assistance. Thus, international disaster management refers to disaster as a hazard that overwhelms the response capability of an affected community. As stated by the UN, international disaster management considers disaster "[a] serious disruption of the functioning of a community or a society causing widespread human, material, economic or environmental losses which exceed the ability of the affected community or society to cope using its own resources."[5] Consequently, the international disaster management community involves international organizations, international financial organizations, regional organizations and agencies, nonprofit organizations, business and industry organizations, local and regional donors, the government(s) of the affected country/countries, governments of aid and donor countries, national emergency management agencies, and the affected population (Borton 1993, 188).

Central to international disaster management is the concept of humanitarian aid regime (Bueno de Mesquita 2007). Humanitarian aid regime is defined as "sets of implicit or explicit principles, norms, rules and decision-making procedures around which actors' expectations converge in a given area of international relations" (Krasner 1983, 2), through which humanitarian aid actors (NGOs, donors, national governments, INGOs, etc.) interact and engage. The explicit objective of humanitarian aid regime is to meet human needs. Within the humanitarian discourse, such objective is conceptualized in terms of the moral obligation to relieve human suffering (Calhoun, 2008, Rieff 2002).

Humanitarian aid is defined in the Preamble to the Statutes of the International Red Cross and Red Crescent Movement, as an aid "to relieve the suffering of individuals, solely guided by their needs," without consideration of other criteria such as "nationality, race, religious beliefs, class or political opinions"—and to "give priority to the most urgent cases of distress" (International Committee of the Red Cross (ICRC) 1986) Catholic Relief Services (U.S.) outlines its mission as that of helping the "impoverished and disadvantaged . . . based solely on need, regardless of their race, religion or ethnicity" (CRS, 2007). CARE USA extends its underlying goal to serve developmental goals for the sake of "the poorest communities in the world," which emphasizes that the function of international development is just as much a function of emergency management. One of the key distinctions that should be drawn between these two definitions lies in the underlying normative assumption of development; that is, the "root causes" of human suffering, which CARE USA seeks to achieve rather than alleviating suffering in the short term.

In this context the term "humanitarian aid" is used to legitimize the party that declares its actions to be "humanitarian" as moral and political concern for human welfare, embracing a politically conscious aid strategy to achieve good outcomes (de Waal 2010, Fassin 2010, Rieff 2002, Rubenstein 2007, 2008, 2014, 2015, Slim 2013, Terry 2002) For that, what counts as a "good outcome" in the highly non-ideal contexts in which international humanitarian aid organizations operate is likely to carry intrinsic normative assumptions. According to Sudanese-born anthropologist Amal Hassan Fadlalla, humanitarian organizations by definition cannot remain neutral: "Humanitarian provision is embedded in broader political agendas, hierarchies and interests that, from the start, render unattainable the notion of impartiality and compromise the wellbeing of the poor and displaced" (Fadlalla 2008).

Fadlalla's argument reflects a growing debate about the definition of humanitarian aid regimes (Eade and Vaux 2007, Smillie and Minear 2004, Bueno de Mesquita 2007, Rubenstein 2007, 2008, 2014, Slim 2013, Terry 2002). In the humanitarian assistance literature, humanitarian aid agencies are often recognized as manifestations of political power or national interests, which may lead to creation of structures that undermine local response and recovery capacities. According to Rubenstein, ". . . while INGO advocates do sometimes engage in representation or act as partners, for the purposes of normative evaluation we should conceptualize INGO advocacy not as representation or partnership, but rather as having and exercising quasigovernmental power. Correspondingly, the main normative standard to which INGO advocates should be held is that they avoid misusing their power" (2014, 208). Bueno de Mesquita provides a logical rationale for the relationship between aid and political power, seeing aid as "an instrument of national policy and as an instrument of humanitarian concerns." (Bueno de Mesquita 2007, 252). Following Bueno de Mesquita's argument, aid delivered from country A to country B creates pro-A policies on behalf of country B, and therefore, aid is conceived as a form of political coercion (2007, 254). Moreover, it is claimed that each agency acts differently to each emergency event, following its own priorities and standards of behavior (Ghani, Lochart, and Carnahan 2005, 11). Winters, for example, suggests that these agencies "have incentives to quickly produce large, identifiable projects rather than to spend costly time harmonizing programming with other donors" (Winters 2012, 2).

The interface between ethics and politics within the humanitarian aid regime becomes more clearly evident in relation to "who gets what, when and how" (Lasswell 1936), which builds on the values, standards, and preferences of each agency in aid allocation (Rubenstein 2007, 2008, 2009, 2014). This problem is intensified in disaster events, when humanitarian aid agencies become bound to their own standards of conduct since they operate outside the areas of established public law. Much of the existing literature that explores these dynamics in the context of emergencies views humanitarian ethics as intangible, highly contextual, not easily visible, and more difficult to codify (Rubenstein 2009, Terry 2002). The impact of agencies' choices and allocative decisions becomes increasingly central to the international disaster management regime, highlighting the interface between ethics and politics. However, the present research suggests that there is a strategic role for the international disaster management community to reconstruct ethics on a global distributive justice foundation, evolving into a tangible and codifiable set of values that could be translated into ethics and professional training programs. If international aid agencies are to meet the needs of disaster-affected populations, the involvement of such organizations within the political process of aid distribution must be accepted as an ethical necessity. Thus, a unified ethical response in international disaster management regime is timely.

The following section considers the institutional efforts made to build shared institutional frameworks for humanitarian bodies engaged in providing assistance in emergency situations.

The Institutionalization of International Disaster Management Regime

Institutional structures of international disaster management systems are created by rules including formal laws, rules, code of conduct, and professional standard, which have become increasingly well-developed and well-established over the past twenty years. However, unlike armed conflicts—where international humanitarian law such as the Third and

Fourth Geneva Conventions, which regulate the provision of food and other goods for prisoners of war, and persons in occupied territories and internees, respectively—natural disasters have no legally binding set of regulations to govern the actions of organizations engaged in humanitarian aid and relief efforts.[6] Reference to natural disaster events in humanitarian law is made only when a natural disaster strikes during the course of an armed-conflict situation. Even the right of humanitarian organizations to offer humanitarian aid to affected states is covered only in armed-conflict situations.[7]

This gap has pushed humanitarian aid organizations and professional emergency management organizations to create codes of conduct and professional standards to regulate and guide their activities in humanitarian aid efforts (Coppola 2011). In 1994 the Red Cross and Red Crescent Movement and several major international NGOs issued a professional Code of Conduct to set out universal standards to govern the activities of relief agencies during disaster events. The Code of Conduct for the International Red Cross and Red Crescent Movement and NGOs in Disaster Response Programs does not employ specific guidelines for operational strategies in delivering humanitarian assistance, but it rather seeks to maintain the high standards of independence, impartiality, and neutrality of humanitarian aid. It includes principles that all NGOs should follow in their disaster response efforts such as impartiality, aid assistance based on needs assessment, neutrality, respect for local culture and custom, building disaster local capacities, reducing future vulnerabilities to disaster, and enhancing accountability.[8] By 2007 more than 400 international and national NGOs had signed the Code of Conduct and, thus, have committed themselves to ensure quality management in humanitarian aid.

The Sphere Project was formed in 1997 by a group of humanitarian NGOs and the Red Cross and Red Crescent Movement. The vision of The Sphere Project is to secure "the right of all people affected by disaster to re-establish their lives . . . and acted upon in ways that respect their voice and promote their dignity, livelihoods and security."[9] The Sphere Project introduces an accountability mechanism to ensure professional conduct by humanitarian actors to their constituents, donors, and effected populations. The Sphere Handbook Humanitarian Charter and Minimum Standards in Humanitarian Response incorporates internationally granted principles and universal minimum standards to guide humanitarian assistance.

In 2001, the Humanitarian Accountability Partnership (HAP) was launched to set guidelines for "making humanitarian action accountable to beneficiaries." The HAP encourages humanitarian agencies to be more accountable to disaster-affected populations through self-regulation, compliance verification, and quality assurance certification. For that, the HAP addressed seven key elements of accountability, e.g., "commitment to humanitarian principles," "capacity-building," "monitoring and reporting compliance," and "communication."[10]

In 2003, the People In Aid, a network of humanitarian assistance agencies, initiated a Code of Good Practice to encourage professional conduct of staff and volunteers engaged in relief and development operations. The "People In Aid Code of Good Practice in the management and support of aid personnel" includes seven key guidelines: "Recruitment and selection"; "Health, safety and security"; "Learning, training and development"; "Consultation and communication"; "Support, management and leadership"; "Staff policies and practices"; and "Human resources strategy." These guidelines are assumed to improve human resources management among humanitarian aid agencies.[11]

The Good Humanitarian Donorship project, which was created in 2003, provides a forum for donors with the aim of facilitating good practice and accountability in funding humanitarian assistance. Such initiatives set out twenty-three guidelines and standards to cope with challenges faced by emergency aid departments in donor governments, such as respect human dignity during and in the aftermath of man-made crises and natural disaster; strive to ensure predictability and flexibility in funding; enhance the flexibility of earmarking and of introducing longer-term funding arrangements.[12]

Despite the evolving codes of conduct, recognized best practices, and formal standards, most of them lack formal enforcement mechanisms to ensure compliance. A serious attempt is made by the International Federation of Red Cross/Red Crescent Societies (IFRC) to create a complete set of International Disaster Response Laws (IDRL). In 2001 the IFRC began its IDRL Program by reviewing studies of international norms, surveys of humanitarian actors, and regional consultations. In November 2007, the IFRC set out the "Guidelines for the domestic facilitation and regulation of international disaster relief and initial recovery assistance" (the "IDRL Guidelines"), which was adopted by all High Contracting Parties to the Geneva Conventions. In 2011, the United Nations Office for the Coordination of Humanitarian Affairs (OCHA), the IFRC, and the Inter-Parliamentary Union conducted the pilot version of their "Model Act for the Facilitation and Regulation of International Disaster Relief and Initial Recovery Assistance" to examine the utilization of the IDRL Guidelines applied in national laws relating to disaster management. The key elements of the IDRL Guidelines include respect for humanity, neutrality, and impartiality. Although these Guidelines are not legally binding nor do they govern interstate relations, they provide a platform for unified legislation across countries in a system that is characterized by different mandates and operating styles such as barriers to entry of goods and people; legal recognition of organizations to operate; and coordination among organizations and governments.[13]

In addition to the IFRC's IDRL Guidelines, the United Nations Office for Disaster Risk Reduction set up the Hyogo Framework for Action (HFA) in 2005. The United Nations International Strategy for Disaster Reduction (UNISDR) is the secretariat of the International Strategy and mandated by the UN General Assembly to ensure its implementation. The UNISDR articulates the objective of humanitarian assistance in international disaster management as "building disaster resilient communities by promoting increased awareness of the importance of disaster reduction as an integral component of sustainable development, with the goal of reducing human, social, economic and environmental losses due to natural hazards and related technological and environmental disasters."[14]

In 2001, the GA, with resolution 56/195, considered that the mandate of UNISDR is to play a key role in the United Nations humanitarian aid system to ensure coordination of disaster reduction and to manage or oversee disaster reduction activities of the United Nations system and regional organizations. In 2005, the Hyogo Framework for Action (HFA) was the first framework document for a common system of coordination to be adopted by Governments around the world.[15]

The HFA outlines strategic goals to achieve disaster resilience of nations and communities to disasters by 2015:

- Enhancement of international cooperation and partnerships

- Applying a multi-dimensional approach to disaster risk reduction in policies, planning, and programming

- Identification of barriers and bias in treating vulnerable persons when planning for disaster risk reduction

- Utilization of culturally sensitive and appropriate interventions based on the gender, race, ethnicity, and age at all levels of disaster risk management policies, plans, and decision-making processes, including risk assessment, early warning, information management, and education and training.

The Hyogo Framework stated key areas that should be tackled such as "(a) Governance: organizational, legal and policy frameworks; (b) Risk identification, assessment, monitoring and early warning; (c) Knowledge management and education; (d) Reducing underlying risk factors; (e) Preparedness for effective response and recovery."

The document concludes with emphasizing the responsibility of States, with the active participation of other actors engaged in risk reduction activities such as local authorities, NGOs, academia, and the private sector. The Hyogo Framework calls for systematic incorporation of risk reduction mechanisms into sustainable development policy, planning, and programming at all levels of regional and international communities, including the international financial institutions, the United Nations System, and the International Strategy for Disaster Reduction (UNISDR).

One of the key pillars for the implementation of the Hyogo Framework for Action (HFA) 2005–2015 is the International Recovery Platform (IRP). The IRP was set as an international source of information exchange on good practice for disaster recovery efforts.[16] Additional initiative to support implementation of the Hyogo Framework for Action (HFA) is the Global Facility for Disaster Reduction and Recovery (GFDRR). The GFDRR is managed by the World Bank to create cooperative activities with other donor organizations to reduce disaster risk and losses. The GFDRR is targeted to enhance the disaster resilience capacity of low- and middle-income countries that are most vulnerable to natural disasters. Following the World Bank criteria, "[l]ow-income countries" receive assistance from the International Development Association (IDA) and "[m]iddle-income countries" receive assistance from the International Bank for Reconstruction and Development (IBRD, together with IDA and the World Bank).[17]

Similar mechanisms to create and sustain the institution of international humanitarian aid in disaster events are used by international professional organizations for emergency managers, such as the International Association of Emergency Managers and the International Emergency Management Society.

The International Association of Emergency Managers (IAEM) is an international organization that aims to promote the goals of reducing human, economic, and social losses due to natural disasters or emergencies. IAEM funds the Certified Emergency Manager and Associate Emergency Manager (AEM) Program to enhance professional behavior among individual emergency managers. The Certified Emergency Manager designation is a nationally and internationally recognized professional certification for emergency managers.

IAEM has issued a Code of Professional Conduct that addresses a range of issues that impact the emergency management professional conduct. The Code aims to increase public trust and confidence in the emergency services provided by members of the IAEM. In addition, the Code is directed at increasing professional competence and ethical behavior. The Code outlines three key principles of respect, commitment, and professionalism. The principle of respect stresses the need to respect supervising officials, colleagues, associates, and aid recipients. The principle of commitment calls for fostering honest and trustworthy relationships, and enhancing stewardship of resources. By professionalism, the IAEM addresses the need to actively promote professional conduct to ensure public confidence and the reputation of emergency management practitioners.

The International Emergency Management Society (TIEMS), registered in Belgium, is another international non-profit NGO. TIEMS serves as a Global Forum for Education, Training, Certification, and Policy in Emergency and Disaster Management. TIEMS's objective is to develop and employ modern emergency management tools and techniques into disaster management practice through information technologies. For that, TIEMS provides a forum for policy guidance to government agencies, industry leaders, academics, volunteer organizations, and other emergency management experts regarding the management of emergencies and disasters. The TIEMS Board of Directors has developed and approved a Code of Conduct for TIEMS. The Code guidelines include protection from discrimination with respect to nationality, race, or creed of any TIEMS member or outside partner, ensure that compensation is disclosed to the TIEMS membership in the annual report to the General Assembly, and the duty of members to report to the TIEMS board about any offer for participation in paid research or similar projects they receive.[18]

Within the European context, in 2001 the European Union (EU) adopted the Community Mechanism for Civil Protection. The Mechanism aims at mobilizing resources and services at the outbreak of disasters requiring urgent response. The Mechanism was created by the European Commission's Directorate-General for Humanitarian Aid & Civil Protection.[19]

The Mechanism operates in a way such that any country inside or outside the Union affected by an intense disaster can make an appeal for assistance through the Emergency Response Coordination Centre (ERCC), Common Emergency and Information System (CECIS), and Civil protection modules that play a coordination role. The ERCC maintains coordination amongst all the participating states in disaster response efforts by pooling the civil protection capabilities of the participating states and maintaining communication channels for useful and updated information on disaster response activities. The Common Emergency and Information System (CECIS) acts as an updated web-based alert and notification designed to provide disaster risk and need assessments. A training program is also part of the Mechanism, which aims at improving the coordination of civil protection assistance interventions. This program involves training courses, joint exercises, and a system of exchange of experts of the participating states to share best practices. Civil protection modules are also mechanisms to facilitate providing national resources from one or more Member States on a voluntary basis.

Drawing on the brief overview of the institutional aspect of the international disaster management system supports the view that the rules, norms, best practices, professional guidelines, and Codes of Ethics/Conduct become a management tool to direct international aid practitioners' ethical obligations in disaster response. Despite differences in institutional structures, standards, and operational strategies, international humanitarian aid organizations share the responsibility of fulfilling the humanitarian needs of the vulnerable communities they serve. As such, the role of international aid actors relies heavily on interventions based on need assessment. In other words, the humanitarian aid activities have implications for aid recipients: who does and does not receive humanitarian aid, which aid services or resources they get, how much, for how long, in what ways resources and services are distributed, and with what unintended implications. For that, the decision-making process held by international humanitarian aid organizations foregrounds issues of distributive justice. The underlying assumption behind the apparent link between institutional structure and global distributive justice is by providing immediate relief; the system of international humanitarian aid ignores the likelihood that later stages of the disaster will affect aid recipients' lives. The objective of international humanitarian aid to alleviate suffering makes it a normative framework; it comes with a commitment to consider unintended consequences of international aid distribution. These assumptions led to the emergence of what is termed the "new humanitarianism." The new humanitarianism is " 'principled,' 'human rights based,' politically sensitive" (Fox 2001, 275). Responding to human suffering with links to human rights and broader political issues is addressed by the Catholic Relief Services: "When considered through the justice/human rights lens, the mere provision of foodstuffs or medical support is an insufficient response to a humanitarian crisis" (Fox 2001, 278). Viewed in this way, new humanitarianism goes beyond the immediate relief of suffering and engages in capacity building, development assistance, and finding long-term solutions to the causes of suffering. Such integration of relief actions and normative discourse generates a tension within new humanitarianism.

In responding to such challenges, the new humanitarianism draws heavily on the discourse of human rights in order to resolve this tension. It refers to the relationship between individuals and their states, and therefore directs humanitarian aid toward protection of human rights. According to Slim: "Rights dignify rather than victimize or patronize people, they make people more powerful as rightful claimants rather than unfortunate beggars. As rights bearers, vulnerable individuals claim for relief assistance as part of their rights as humans. Rights reveal all people as moral political and legal equals" (Slim 2002, 16). The emphasis of human rights doctrine leads to viewing international humanitarian aid as part of a political (universal) project to transform the world into a better one in which human rights are realized and protected. This imperative is entrenched in both the international humanitarian law (IHL) and the international human rights law (IHRL). The IHL is a set of international rules created by custom or treaty that addresses humanitarian problems arising from international or non-international armed conflicts. The IHRL is also a set of guidelines established by treaty or custom intended to uphold and protect human rights at the international, regional, and domestic levels. Both laws aim to ensure that the lives, health, and dignity of individuals will be protected, while the rules of IHL deal with issues that fall outside the purview of IHRL, such as the conduct of hostilities, combatant and prisoner of war status, and the protection of the Red Cross and Red Crescent emblems. Similarly, IHRL deals with aspects of life in "normal times" that are not regulated by IHL, such as freedom of the press, and the right of assembly, to vote, and to strike. This institutional framework defends human rights by providing a moral foundation and a set of standards to guide international humanitarian aid practitioners.

Although the human rights approach is offered as an answer to questions of distributive justice, difficulties also arise within the rights-based approach concerning the call for a universal community of justice that challenges that state's claim to exclusive national sovereignty over its people. Thus, the decisions shaped in part by the institutional structure of international humanitarian aid in defense of human rights represents a political as well as moral intervention because it is a claim to constrain and hinder state activity. In addition, the human rights approach to humanitarian aid provision ignores the difference between what counts as just during or after natural disasters and what counts as just under circumstances of normal times. For example, since aid resources are scarce, wasting them leads directly to fewer lives saved; cost-effectiveness and efficiency judgments might gain priority over other social justice values such as the efforts to ensure equality of resources or equality of outcomes. By broadening the scope of humanitarian aid provision to encompass the unintended impacts of international aid allocation decisions, ethical responsibility must be reformulated to take into account global distributive justice demands.

In disaster events, when aid recipients are viewed as vulnerable, international humanitarian aid is understood as possessing legitimate authority over resource allocation. In a way, the practice of humanitarian aid is likely to generate a sense of superiority on the part of the aid providers, who are in a position to supply immediate resources and services for relief. The recipient is apparently incapable of relying on his own capacities and therefore the superior aid provider should be beneficent toward him. This practice creates conditions for inequality and power relations. Thus, humanitarian aid "undermines the idea that people are the subjects of their own survival and [of] equal worth to their benefactors" (Slim 2002, 6). The implications of this aspect of international humanitarian aid practice are that aid providers gain the power to decide what "counts" as an emergency as well as to make use of the "windows of opportunity" created by disasters to promote longer-term objectives such as how to help communities become more resilient in the face of future disasters. Consequently, the humanitarian aid system may influence the way need is assessed and measured.

The institutional structure of international humanitarian aid systems constitutes the process by which humanitarian aid organizations make allocative decisions that affect the life-prospects of people and nations receiving humanitarian aid. For that, the institution of international humanitarian aid is necessarily required to mediate the relationship between universalizing and particularizing practices. While the right-based discourse refers to a universal duty to assist those in need, the global distributive justice discourse incorporates the understanding of our associational connectivity in a complex and globalized world and the justification of ends and means by which such responsibility should be enacted. Distributive justice discourse points directly to how universal duty to assist should be practiced: that is, on the rules and norms that shape the distribution of aid resources and services by the international humanitarian aid system as a whole. In the disaster management context, international humanitarian aid actors have multiple principals whose interests and priorities are multiple as well. The principal that is foundational to the legitimacy of international humanitarian aid is social justice itself. The duty to assist people in emergency situations addressed to the institution of international humanitarian aid steps beyond the realm of new humanitarianism by making the members of such institutions moral agents embedded in a complex structure of global interconnection. Although the distributive justice discourse is imbued with power by engaging in decisions of who is accountable for what and when, it is the ethical setting most conducive to and predictive of responsible moral conduct and therefore meant to be strikingly depoliticized in its application. This highlights how the ethics of global relational embeddedness reformulates the ethical responsibilities of international aid actors in ways that raise their awareness and sensitivity of ethical dilemmas and how international aid actors can reclaim control over their professional evolution.

By drawing on what ethical responsibilities the international humanitarian aid community has in times of disaster to make the institution of international humanitarian aid more just, ethical dilemmas related to distributive justice come to the fore. In the following chapter we address the dilemmas of international humanitarian aid provision arising from aid allocation in disaster events.

2

Dilemmas and Ethical Issues in International Disaster Management

> The notion that "being humanitarian" and "doing good" are somehow inevitably the same is a hard one to shake off. For many people, it is almost counter-intuitive to have to consider that humanitarian action may also have a dark side which compromises as well as helps the people whose suffering it seeks to assuage.
>
> —Slim 1997, 244

Large scale disasters overwhelm the capacity of affected governments to meet the needs of their citizens (Adams and Balfour 2009, Gadsen 2008, Ink 2006, Jia 2008, Menzel 2006, Mooney 2009, Yang 2008). When states are unable to meet citizens' needs, response efforts are centered on the international aid community to quickly mobilize response resources and assist affected populations. The accountability held by humanitarian aid organizations in a disaster setting is conceived in terms of supporting vulnerable communities and helping vulnerable citizens and groups articulate and meet their shared interests and needs (Aldrich and Crook 2008, Ink 2006, Menzel 2006, Nickel and Eikenberry 2007, Özerdem and Jacoby 2006, Stivers 2007, Ackerman 2005, Koliba, Meek, and Zia 2010, Rodan and Hughes 2012, Smith 2008). Recently, at the outbreak of the powerful Typhoon Haiyan (also known as Typhoon Yolanda) the Philippines government failed to address disaster relief and coordination among its administrative agencies, whereas two days after the powerful typhoon, international organizations mobilized a wider, more effective international collaborative system.[1]

The effectiveness of the international disaster management community depends heavily on the willingness of disaster affected government to accept international aid and cooperate with international aid actors.

Disaster-affected governments and public administration have criticized international disaster response and relief efforts by arguing that humanitarian aid organizations may use the window of opportunity generated by natural disaster disruption to extend their agendas and provide strong criticism over the priorities and methods of the elite-controlled political processes (Özerdem and Jacoby 2006, Forgette, King, and Dettrey 2008).

This book does not attempt to validate such claims. Instead, it suggests that if international disaster management is to be pursued as professional, a systematic approach to articulate ethical responsibilities and core values of the profession is needed to tackle these perceptions and build greater trust between international aid actors and disaster-affected government agencies. As an expression of their ethical responsibility, international aid actors must communicate their concerns and ethical/moral dilemmas faced by international humanitarian aid organizations that beset disaster response and relief efforts and hampered the development of effective aid delivery systems. Drawing on ethical and professional dilemmas related to international disaster management helps to translate the humanitarian aid mission into an operative disaster management system involving a series of distributive justice decisions about who should receive aid, what types of aid resources and services are needed, and how these resources should be allocated. This process follows the institutionalization process of international humanitarian aid, as explained in the previous chapter. That is, the ability of an agency to successfully resolve the following dilemmas will determine the nature and quality of international disaster management. Specifically, we address six basic dilemmas encountered by the international disaster management community: (1) the dependency syndrome, (2) donation fatigue, (3) discrimination, (4) corruption, (5) the relationship of disaster management to ongoing development processes, and (6) climate change compensation. A useful starting point for the analysis of ethical dilemmas related to international disaster management is to clarify the differences between ethical dilemmas and ethical issues, with reference to the manner in which ethical theory may contribute to ethical decision making.

Ethical/Moral Dilemmas vs. Moral Issues

Before we get started, a short note on the use of the terms "ethics" and "morality" is required. Like most philosophers, in this book we prefer to use these terms interchangeably. As a practical matter, the distinction between the terms ethics and morality is not always clear. Historically, the term *ethics* comes from the Greek word *ethos* meaning customs and habits of individuals. Morality is rooted in the Latin word *moris*, which indicates basically the same thing. Those who draw a clear distinction between them use the term *morality* to denote personal values while *ethics* refers to social values reflected in rules and standards of forms of living together such as groups, organizations, or mankind (Jeong and Han 2013).

In our daily lives, we are faced with moral and ethical decision making. Ethical and moral dilemmas are meant to denote symmetrical dilemmas that involve not decisions between right and wrong, but between right and right. In other words, ethical/moral dilemmas involve no morally relevant difference between two possible courses of action or between two or more values in the perception of a decision maker[2] (Conee 1989, Foot 1975, Gowans 1994, Ross 1930, Sinnot-Armstrong 1988, Williams 1973). According to Joseph Badaracco, "We have all experienced situations in which our professional responsibilities unexpectedly come into conflict with our deepest values, we are caught in a conflict between right and right. And no matter which option we choose, we feel like we've come up short." In contrast, ethical or moral issues arise when single values or concerns, such as fairness, transparency, integrity, and effectiveness are included in the decision process. Moral issues are defined in situations where it seems fairly obvious to most people what is or was the right or wrong course of action.[3]

One of the classical examples for moral dilemma is represented in the novel entitled *Sophie's Choice* by William Styron (1980). Sophie Zawistowska, a Polish woman, is arrested by the Nazis and taken to the Auschwitz concentration camp with her two children. On arrival, since she is not a Jew, she has the opportunity to choose one of her children who will then be killed by the gas chamber, whereas the other child will survive. In case of indecision, both of her children will be killed. When faced with this dilemma she chose to let the Nazis take her daughter, who is younger and smaller, for her son, older and stronger, thus has better chances to survive. Years later, Sophie is unable to bear the consequences of her choice and that is why she commits suicide. Ruth Barcan Marcus (1980) poses a similar dilemma in which an agent must choose which twin to save over the other. In this dilemma, Marcus shows that for reasons of time or location constraints both options have comparable moral support, but both cannot be performed—a disjunctive obligation. Another example of moral dilemma is offered by Jean-Paul Sartre in his essay "Existentialism Is a Humanism" (1946). Sartre brought a dilemma posed by his student: "His father was on bad terms with his mother, and, moreover was inclined to be a collaborationist; his older brother had been killed in the German offensive of 1940, and the young man, with somewhat immature but generous feelings, wanted to avenge him. His mother lived alone with him, very much upset by the half-treason of her husband and the death of her older son; the boy was her only consolation.

The boy was faced with the choice of leaving for England and joining the Free French Forces—that is, leaving his mother behind—or remaining with his mother and helping her to carry on.

He was fully aware that the woman lived only for him and that his going-off—and perhaps his death—would plunge her into despair. He was also aware that every act he did for his mother's sake was a sure thing, in the sense that he was helping her to carry on; whereas, every effort he made toward going off and fighting was an uncertain move that might run aground and prove completely useless. For example, on his way to England he might, while passing through Spain, be detained indefinitely in a Spanish camp; he might reach England or Algiers and be stuck in an office at a desk job. As a result, he was faced with two very different kinds of action: one, concrete, immediate, but concerning only one individual; the other concerned with an incomparably vaster group, a national collectivity, but for that reason was dubious, and might be interrupted *en route*. . . . He had to choose between the two . . ." (Sartre [1946] (2002), 447). The way Sartre presents the situation is a moral dilemma in which the student is forced to choose between conflicting moral imperatives.

Characteristics of Moral Dilemmas and Moral Obligations

Drawing on these examples of moral dilemmas, we can identify the characteristics of a moral dilemma. A moral dilemma is a situation in which we are faced with two moral obligations that conflict because we are unable to hold both obligations together due to contingent circumstances.

A moral dilemma is then described as follows:

1. There is an obligation to do A.

2. There is an obligation to do B.

3. Alternatives A and B are mutually exclusive act-choices.

4. The given circumstances make it impossible to perform both alternatives A and B.

In the case of Sophie's choice and Marcus's twin dilemma, the agents have an obligation to save one child while at the same time having an obligation to save the other child. Alternatively, we can express it in a negative manner by showing that the agents are faced with an obligation *not* to let one child die, but they are also faced with an obligation *not* to let the other child die.

In Sartre's student's dilemma, the point of view of the agent himself seems to be crucial to identifying the situation as a moral dilemma, that is, the decision between two morally compelling choices. We might not identify the student's situation as a moral dilemma in that other agents may not view these two options as mutually exclusive.

The subjective aspect of moral dilemmas leads us to draw on another feature of moral dilemma, namely, the psychological force of dilemmas. Moral dilemmas may cause psychological pain for the agent who is forced to make the difficult moral choice.

Moral issues for instance can be solved through deliberation by using a rational calculation mechanism in order to assess the existing options and their apparent consequences, yet for a moral dilemma one should go beyond deliberation as the agent faces physical and psychological stress at *having to decide* between two mutually exclusive courses of action, both of which are ethically justifiable, and neither of which is optimal (Harding 1985).

Finally, a moral dilemma produces a final choice that is merely *morally maximal*. Since the agent is obliged to make a choice that he ordinarily would not wish to make, she or he is left with an option that offers the maximal expected utility. In moral dilemma, the circumstances constrain the ability to make the morally optimal choice (choosing both or neither). For that, the agent must reconcile following through with the perceived maximum of expected utility if the agent forgoes removing one of the options.

Ethical Decision Making/Moral Judgment

Moral dilemmas involve weighing the ethical justification for alternative courses of action. Cognitive theories of moral development stress the need to draw on cognitive capacities that affect an individual's moral consciousness. According to Kohlberg's (1969) six-stage model, moral development is actualized through the capacity of an individual to solve a moral dilemma through maximally balanced and autonomous decision making. Thus, a person's moral reasoning may continue to evolve toward a higher, more principled level of reasoning as a result of cognitive maturity, experience, education, and appropriate environment (Kohlberg 1981). Despite criticism over Kohlberg's stagist model of moral development (Gibbs 1979, Gilligan 1982, Habermas 1979, Snell 1996, Sullivan 1977, White 1999) regarding the rooted Western cultural bias and gender bias in the model, the excessive focus on cognitive aspects of moral reasoning and the fact that emotional and affective aspects are ignored, Kohlberg's model remains perhaps one of the most distinguished theories of moral development.

The educational objectives underlying the use of moral dilemmas can be summarized as follows: (a) to develop maturity of the person's moral judgment structures, in both complexity and depth, as manifested in a fair and balanced resolution of a moral dilemma; (b) to enhance the person's critical self-reflection when identifying and resolving moral dilemmas to develop a sense of autonomy; (c) to provide the person with more common and intense ethical questions, both social and professional; (d) to advance greater sensitivity to moral dilemmas arising in their daily practice that may be ignored as part of a decision making; and (e) to promote ethical leadership to foster virtuous organizational and professional culture.

Ethical Decision-Making

The ethical decision-making process in which individuals make ethical decisions consists of four stages as addressed by Rest's (1986) model (see Figure 2.1.) The four stages correspond to distinct psychological processes including moral sensitivity, moral judgment, moral motivation/intention, and moral character/action.

| Identification of ethical dilemma | Ethical Judgment | Intention to act ethically | Ethical action/behavior |

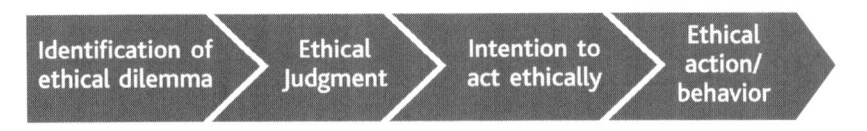

Figure 2.1. Rest's (1986) four stages model of ethical deliberation.

Following Rest's model, there is a four-step process for solving an ethical dilemma:

1. **Stage One: Identification of an Ethical Dilemma.** This stage is highly important as it involves a moral awareness that a dilemma may affect the well-being of others. Later attempts were made to broaden the definition of moral awareness, suggesting that moral awareness involves a person's sensitivity that a situation incorporates moral content that validates a moral perspective (Reynolds 2006). This definition points to a genuine moral concern for other persons. According to Chia and Mee (2000, 255), "When individuals recognize the moral dimension of an issue, this recognition has the potential to influence their judgments, intentions and decisions."

2. **Stage Two: Making an Ethical Judgment.** The realization of a situation as a moral dilemma entails an assessment of the expected outcomes to occur in a given situation. Ethical judgment is constituted in part by a person's moral competency, which is associated with Kohlberg's (1969) highest stage of moral development. Normative ethical theory is meant to support ethical judgment by identifying relevant features that help to tip the balance one way or the other in moral judgment. This requires moral reasoning through the possible choices and impending consequences to determine which are ethically sound.

3. **Stage Three: Intension to Act Ethically.** This stage is based on a person's commitment to act ethically to resolve a moral dilemma. According to Fishbein and Ajzen (1975), intensions play a critical role in linking ethical judgment to action in the ethical decision-making process (Barnett 2001, Singhapakdi, Vitell, and Kraft 1996). These intensions, when referring to moral motivation, are the person's intentions to make the morally optimal choice(s) and to do so.

4. **Stage Four: Ethical Action/Behavior.** This stage refers to implementation of the ethical action. It involves the person's ability to exhibit moral courage (moral action); that is, to follow through with the moral decision (Jones, Massey, and Thorne 2003). Rest's model may present a simplistic link between moral judgment and action while critics have shown that moral action develops independently of moral judgment (Brown and Hernstein 1975) and that individuals positioned in constraining environments may act immorally despite their capacity for moral reasoning (White 1999).

Ethical Dilemmas in International Disaster Management

As seen in the previous chapter, many international humanitarian aid organizations engaged in disaster response and relief efforts have comprehensive Codes of Conduct or Codes of Ethics, updating rules and regulations. However, in this book we are concerned with international humanitarian aid organizations as critical players within environments that they cannot fully control, such as disaster events. In these situations, ethical dilemmas arise from the unintended consequences of response efforts held by international aid actors. International humanitarian aid practitioners may find it hard to function independently, relying on autonomous governance structures; in disaster events they need to collaborate with different types of actors such as other INGOs, local agencies, and governments for the sake of providing immediate aid and reaching optimal relief outcomes. In such context, international disaster management community ethical decision making should be guided by shared values to meet the challenges of the disaster setting.

Moral dilemmas emerge when the international humanitarian aid organizations know the right course of action based on their organizational or professional rules but may choose to act differently because of institutional obstacles such as lack of collaboration, lack of proper information on needs and risks assessment, hierarchical power structures, lack of resources, lack of support, or legal limits. In these instances, disaster management needs to incorporate ethics to foster members' commitment to adhere to professional rules and virtues and to establish infrastructure to ensure that ethics is embedded into international disaster management practices. It is assumed that a Code of Ethics has the capacity to universalize professional understandings of ethics in international disaster management and operationalize core values and responsibilities. For that, a Code of Ethics is intended to be of assistance to international aid actors when faced with ethical dilemmas arise in aid allocation decisions.

Although we do not claim to present an exhaustive list of the ethical dilemmas encountered by the international disaster management community, the purpose of the subsequent chapters is to identify ethical dilemmas associated with the allocation of relief and response resources and services that affect the life-prospects of people and nations receiving humanitarian aid. We identify four kinds of ethical dilemmas frequently faced by the international disaster management community that reflect the distributive concerns in relation to who does and does not receive humanitarian aid, which services or resources of aid they get, for how long, in what ways resources and services are distributed.

1. The Dependency Syndrome

The ethical dilemma of dependency syndrome refers to the "when" aspect in distributive decision making. It bears on the time limits that are adopted out of concern for continued reliance on international aid resources and services. The mandate of humanitarian aid and relief organizations should be to supplement affected national governments and their available resources. We use the term *syndrome* to capture the way disaster relief and response aid manifest as a coincident pattern associated with ambiguous social conditions. Viewed in this way, dependency syndrome is perceived as a cluster of negative aspects of humanitarian relief and something to be avoided. Central to these perceptions is that reliance on assistance being provided would be a bad thing, creating disincentive effects on recipient societies that undermine community- or individual-based initiatives or competencies. Dependency syndrome then results from the reliance of the disaster-affected population (receivers) on humanitarian aid and on its providers. The ethical dilemma lies in the confusion between the aid agency's assessment of affected communities' subsistence needs, and those communities' perception of the role of the aid agency. This can sometimes lead to continuing reliance on international aid actors to meet citizens' needs while ignoring or sidelineing state response and relief capacities. The need for continued provision of relief is seen as the problem, rather than a symptom of poverty, destitution, or local conflict that intensified in extreme events. Given that the goals of humanitarian aid are to alleviate suffering in the face of adversity, it is suggested that emergency relief should be distributed only when national governments' capacities to respond are overwhelmed by a disaster event. This initiates a full assessment of the needs and the capacities of the affected population, which increases in difficulty and complexity with the size and scope of the disaster.

2. Donation Fatigue

The ethical dilemma of donation fatigue deals with the "what" aspect in distributive decision making to ensure appropriate disaster response funding for disaster affected governments. Despite the growing number and size of natural disaster events, several countries are already falling short on their aid obligations. Strong and stable support for foreign aid among taxpayers and voters is one of the key mechanisms for disaster-affected governments to cope with the costs of disaster and to establish long-term disaster risk reduction management plans. The term "donation fatigue" is used here in a broader sense to denote how international aid agencies' opinions of recipient countries may justify weakening foreign aid. Donor fatigue is caused by budget exhaustion. It may wipe out the donation funding raised by humanitarian aid and relief organizations. In other instances, humanitarian aid and relief organizations may grow frustrated with constant appeals for donations. Frustration may also be intensified when the international community donates to national governments that mismanage loans and donations, resulting in political and administrative corruption and/or underdevelopment coupled with high indebtedness. Despite the efforts of cosmopolitan justice to push for global aid involvement, continual mitigation of large-scale disaster events leads to fatigue and makes it much more difficult to mobilize global public sentiment in the way necessary to overcome the desires for lessening support for foreign aid.

3. Corruption

The ethical dilemma of corruption refers to the "how" aspect in distributive decision making in addressing ways to overcome misuse of international aid and to ensure that it finds its way to those in need. Humanitarian aid and assistance involve large amounts of cash and supplies that may create opportunities for corruption, especially in poor countries. The problem, therefore, is not corruption per se, but the way aid flows is creating additional incentives for corruption within recipient countries; for aid agencies, existing corruption patterns may justify declining aid flows from the world's rich nations, which in turn may result in developing countries' economic stagnation. The allocation of aid then undermines the moral obligation behind humanitarian assistance, leading to inequitable and ineffective distribution of disaster relief and reconstruction aid.

One of the predictable consequences under these circumstances is that poor people's subsistence needs are marginalized and development outcomes suffer in the long run. During disaster response and relief efforts, citizens are more prone to recognize that public resources are not being managed effectively and that corruption in the use of aid relief undermines immediate relief efforts.

4. Compensation

The ethical dilemma of compensation deals with the "who" aspect in distributive decision making with regard to who should be compensated for bearing climate-related damages. The recent devastating typhoon that caused the death of thousands of people in the Philippines has reinvigorated the debate about whether rich nations, which contribute to global warming, should compensate poor ones for climate-related damages. The issue of climate change loss and damage compensation was recently raised at United Nations climate talks in Warsaw in November 2013. Philippines President Benigno Aquino III urged countries to take moral responsibility and assure that the developing world is compensated to the level that vulnerable countries and communities will become more resilient. Climate change raises a wide range of challenging ethical questions, including the fact that the harm to humans is increasingly born by people who have contributed little to greenhouse gas emissions, namely populations in the least developed countries and future generations. Given the severity of disaster costs and damages, it is crucial for the international community to address the issue of compensation that would lead to fair distribution of responsibilities to reduce greenhouse gas emissions and support adaptation programs of individuals, communities, and states harmed by climate change.

All these ethical dilemmas touch upon the unresolved question of whether disaster aid is an end of itself and should be maintained as an indispensable or rather "helpful means." The economist Joseph Stiglitz refers to this unsettled controversy of demarcating a line between ends and means as one that continues at the level of policymaking:

> There are important disagreements about economic and social policy in our democracies. Some of these disagreements are about values—how concerned should we be about our environment (how much environmental degradation should we tolerate, if it allows us to have a higher GDP); how concerned should we be about the poor (how much sacrifice in our total income should we be willing to make, if it allows some of the poor to move out of poverty, or to be slightly better off); or how concerned should we be about democracy (are we willing to compromise on basic rights, such as rights to association, if we believe that as a result, the economy will grow faster). (Stiglitz 2002, 218–19)

Our goal in the book is both analytical and prescriptive. Unidentified and unresolved ethical dilemmas in a disaster setting can lead to feelings of uncertainty; tension with other aid agencies, governments, and affected communities; and frustration in disaster management. Today's international disaster management environment, with its shortage of staff and resources as well as economic constraints, can be overwhelming for international aid organizations who try to uphold their aid duties and ethical and professional principles. It is suggested that ethical dilemmas associated with disaster aid resource allocation have to be defined in ways that spell out their potential for translating general ethical principles and values into action (agency) by drawing on different approaches to distributive justice that have different advantages and disadvantages which vary in importance from context to context, and any optimal solution must bear that in mind.

The normative function of international disaster management structure in this manner is made up of most basic principles, values, and norms that can serve as a guide to allocation. In disaster events, disputes over allocations are inevitable, and characterize communal living, cutting across cultural as well as historical parameters. Resource allocation is not only unavoidable but is a minimum prerequisite of any organization that engages in disaster aid provision. Given the frequency with which international aid organizations encounter distribution problems, and the consequences of allocative decisions, identifying and examining the principles such international disaster management communities use to guide their distribution decisions is an important area of inquiry. Numerous principles may be used by people distributing resources among group members. Resources might be distributed according to the needs of the potential recipients, or in accordance with the contribution of each group member, or just split equally among all recipients. Thus, access to such resources will reflect the rules and norms that govern distribution and exchange in different institutional arenas (Leventhal 1976). These rules and norms empower certain actors to dictate the principles of distribution and exchange so that the distribution of resources tends to be embedded within the distribution of authoritative resources (Giddens 1979)—the capacity to define priorities and enforce claims.

Studies conducted on distributive and procedural justice have focused on people's perceptions of fair treatment or receiving fair outcomes through authoritative allocative outcomes (Lind and Tyler 1988). These studies suggest that the exercise of authority should take into account public satisfaction with decisions to enhance public participation efforts. Following in a merely instrumental way, while still leaving citizens feeling they were treated unfairly, citizen engagement efforts will decrease and so will their trust in authorities. The question is: How do people assess the fairness of a decision? Social psychology theories and research suggest that both the decision outcome and the process by which that decision is reached affect perceptions of justice (Hegtvedt and Markovsky 1995). The theories also provide a clear set of principles people use, i.e., resources, when assessing the fairness of the process and outcome of decisions.

Scholars from various disciplines have addressed the conceptualization of distributive justice judgment—that is, evaluation of how social resources should be allocated, whereby individuals or groups feel that they receive what they deserve (Bierhoff, Cohen, and Greenberg 1986, Cook and Hegtvedt 1983, Deutsch 1985). However, their formulations do not cover every instance of distributive justice and do not expressly consider other dimensions of distributive justice evaluation as relevant criteria for distribution of goods. From this criticism has emerged an analytical approach that is intended to be a method for the analysis of accounts of distributive justice and which conceives distributive justice judgments as multidimensional and contextual.

Specifically, the analysis of ethical dilemmas faced by the international disaster management community will be based on comparative analysis of distributive justice as it relates to five major facets (conceptual classifications) of distributive justice judgments: distributive good, community of justice, preconditions of distribution, the structure of principles and rules, and their content. In the following chapter, we attempt to map their content domains so as to distinguish between political theories of distributive justice. Further, we provide careful weighing of the relative affinities and contrasts of distributive justice judgments for dealing with ethical dilemmas that may enable international humanitarian aid organizations engaged in disaster response and relief efforts to reflect on new problems and principles that challenge their existing institutional frameworks.

3

Global Distributive Justice

> Anyone can . . . give or spend money. But it is not easy to decide to whom to give how much, when, for what purpose, and how.
>
> —Aristotle, *Nicomachean Ethic*, bk. II, chap. 9

This chapter attempts to bridge the previous chapter and subsequent chapters by developing a comprehensive framework for ethical decision making in international disaster management. In this chapter we seek to refine the sense of the duty of assistance and expand it to consider the long-term consequences of the allocative decisions and actions made by humanitarian aid organizations. An extended sense of the duty of assistance is embedded in contemporary cosmopolitan justice. This chapter provides a backdrop to our inquiry of the long tradition of philosophical reflection on distributive justice, which continues through contemporary dialogue about cosmopolitan distributive justice generated by political philosophers and theorists. It aims to unveil the structure of distributive justice judgments, using a facet approach as a formal framework for expressing the multidimensionality of distributive justice evaluation, including distributive good, community of justice, preconditions of distribution, the structure of principles and rules, and their content. In evaluating the three well-established theories of global justice including the consequence-oriented approach developed by Peter Singer, the rights- and institutions-oriented approach developed by Thomas Pogge, and the capabilities-oriented approach of Amartya Sen, this chapter questions which global justice approach has the potential to meet distributive justice criteria.

The Nature of Distributive Justice

Distributive justice is generally associated with the goal of alleviating economic deprivation and with the allocative effects of policymaking and social reform. In this chapter we are concerned with distributive justice, which supposes that individuals and communities within society are bound in the distribution of potential benefits and potential burdens as a collaborative arrangement. As societies have limited wealth and resources, a question must be raised as to how those benefits ought to be distributed among the members of a given society. Fairness is a common guideline in allocation of public assets so that each individual receives a "fair share." But this leaves open the question of what constitutes a "fair share." Thus distributive justice can be very broadly defined as concerned with the fair allocation of resources among diverse members of a community. A fair allocation system typically takes into consideration the total quantity of goods to be distributed, the distribution procedure, and the pattern of distribution that follows.

In the philosophical literature, distributive justice has been studied in prescriptive terms, that is, in terms of defensible arguments that justify how different resources ought to be distributed in society. As phrased by Roemer, "The theory of distributive justice—how a society or group should allocate its scarce resources or product among individuals with competing needs or claims—goes back at least two millennia. Aristotle and Plato wrote on the question, and the Talmud recommends solutions to the distribution of an estate among the deceased's creditors" (Roemer 1993, 1).

Aristotle, Plato, and the Talmud, like other ancient legal texts, did not discuss distributive justice in its modern sense, but rather societal principles or resource allocation. Indeed, the sense of distributive justice as found in Aristotle's writing referred to the principles ensuring that deserving people are rewarded in accordance with their merits, especially regarding their political status (Aristotle 1934). In its modern sense, distributive justice is more than competing claims to property; it is how to allocate scarce resources as a matter of justice that meets everyone's needs (Fleischacker 2004). The fact that certain kinds of people ought to live in need as a part of divine order or lack of merit bridged the ancient philosophy with the modern philosophy of distributive justice. David Hume suggested that man's "wants and necessities" are "numberless," yet his means for satisfying these necessities are "slender" (Hume 1978, 484). According to Hume, it is only through cooperative interaction with his fellows that man can "remedy" his slender means and natural defects, thereby increasing his "force, ability and security" and improve those goods and possessions acquired through "industry and good fortune" (ibid., 487–8). Thus, the primary advantages of man's collaboration with other individuals in society are revealed in the provision of security and welfare and the increased capacities for the production of goods.

The modern concept of distributive justice rests on state authority to enforce a redistribution of property rights throughout society so that everyone is supplied with a certain level of material means (Foa 1971). If the level of goods everyone ought to have is low enough, it may be that the market can guarantee an adequate distribution; if everyone ought to have an ample basket of welfare protections, the state may need to redistribute goods to correct for market imperfections; if what everyone ought to have is an equal share of all goods, private property and market-oriented economies will probably have to be replaced altogether by a state redistribution system. Thus, the very idea of distributive justice plays a protective role in property rights, such that it may even entail a denial of private property.

Review of Distributive Justice Theories

Contemporary distributive justice theory has been influenced, in contrasting ways, by different disciplines including sociology and social psychology, philosophy, and economics. Thus distribution is often conceptualized in terms of social justice judgments, that is, evaluations of social systems in order to understand how people perceive economic and social inequality and what criteria they use in evaluating the just or unjust. (For overviews, see Berger et al. 1983, Bierhoff, Cohen, and Greenberg 1986, Brickman et al. 1981, Cook and Hegtvedt 1983, Deutsch 1985, and Miller 1992.)

To cover a complex term such as distributive justice in a chapter may seem an impossible task. What makes it possible to introduce this vast topic in a meaningful way is the fact that the literature on distributive justice is characterized by two structural parameters: egalitarian and multi-principle approaches.

The predominant approach to studying distributive justice—equity theory—has focused on the tension produced by negative emotions resulting from individual inequity, perceptions that influence individuals to alter their performance and attitudes, while the actual effects of equity perceptions on the receiver's affective position and subsequent behavior have not been fully explored. During the early 1960s, research on equity in distributive justice has focused primarily on wage inequalities, with experimental studies of the relationship between wage inequities, productivity, and work quality (cf. Adams 1963, Adams and Rosenbaum 1962). Equity theory is concerned with negative emotions of employees, such as anger and frustration resulting from their increased perceptions that their actions are unfairly compensated (that is, if distributive rewards are perceived to be allocated unfairly), resulting in a decline in employee performance and cooperation to a point where the perceived inequity is corrected. Thus, equity theory posits a twofold response to perceived unfairness: First, employees experience an emotional reaction (i.e., negative emotion) to perceived unfairness; second, they are motivated by this negative emotional response to change the situation, to reestablish equity (Mowday 1996). The nature of psychological (emotional) reaction to inequity was embedded in cognitive consistency theories (Festinger 1957, Heider 1946, 1958; see also Lewin 1935, 1936); the perception and subsequent emotional reactions of inequity create tension resulting in dissatisfaction, anger, guilt, etc. (Adams 1965, Homans 1958, 1961, 1974). Additional research conducted at about the same time focused on equity in terms of discrepancy (e.g., Adams 1963, 1965, Walster, Berscheid, and Walster 1973, 1978). Discrepancy is defined as inequity or distributive injustice. The conceptualization of discrepancy is derived from a perceived mismatch between inputs and outcomes or as a mismatch between the expected and applied distribution principles, respectively. Based upon Homans's early ideas (1958, 1961), distributive justice obtains when partners in an exchange relationship receive rewards that are proportional to their investments or contributions.

During the late 1960s scholars reexamined the fundamental assumptions of equity theory. Several theorists (e.g., Hegtvedt and Markovsky 1995, Mowday 1996, Opsahl and Dunnette 1966, Wicklund and Brehm 1976) have pointed out that equity theory needs increased accuracy in its predictions of individual responses to inequity. Much of this research reflected a gradual shift in orientation, which was partially a result of the vagueness of the research results derived in these early studies and of the limitations of the Adams experimental situation (Andrews and Valenzi 1970, Lawler 1968, Wiener 1970). These studies include analyses of the effects of such variables as population characteristics including age; sex (Lane and Mess 1971); personality traits, e.g., ascendancy, poise, Machiavellianism, altruism, individualism, authoritarianism (Blumstein and Weinstein 1969, Lane and Mess 1971); and situational determinants, e.g., variations in experimental tasks (Weick 1964, 1966, Wiener 1970); source of the inequity (Leventhal and Michaels 1969); and sufficiency of the total amount of rewards to be distributed (Lane and Mess 1972).

Other scholars criticized the equity canons of distributive justice for being monistic, focusing on one single factor as the basis for a legitimate claim. Following Rescher (1966, 82): "We must take a multifaceted approach to claims because of the propriety of recognizing different claim grounds as appropriate types of distribution." In a similar vein, Blau (1964, 221) and Lenski (1966, 52) addressed the idea of facet analysis of distributive justice evaluation by showing how distributive justice evaluations rest upon fundamentally different legitimate distributive rules that require at least implicit social approval. For example, Blau (1964) showed how a leader who uses power to keep subordinates agreeably subservient will be rewarded with the legitimization of his authority up to a certain degree at which point an oppressive use of power may engender collective reaction and encourage the development of an alternative ideology, i.e., an alternative distribution rule. Lenski (1966, 52) also suggests that building legitimization through laws and propaganda will follow the seizure of power, but that the form of legitimization should be grounded in existing consensual norms. Thus, the equity of distribution rests on different sources of legitimization of or opposition to authority, and by individuals' perceptions that a particular situation or set of circumstances is inequitable.

Presenting distributive justice that resonates with different stratification systems thus allows for multiple sources of inequity or combinations of inequities that have not yet been addressed by earlier equity theorists.

During the 1980s, critics of the equalitarian approach challenged the assumption that the equity principle is universal and that it applies to all types of resources (Törnblom and Vermunt 2007). They claim that equity principle is not inclusive enough to cover every instance of distributive justice and other principles relevant for normative allocative choices (Deutsch 1985, Leventhal 1980, Rubinstein 1988, Sampson 1975). From this criticism emerged the multi-principle approach, which conceives social justice judgment as multidimensional and contingent. Social justice judgments are based on several basic principles of justice from which a variety of sub-principles, known as "distributive rules," are derived. These distributive rules are applicable to different types of decisions in judging distribution as fair (Galston 1980, Walzer 1983) in different situations (Leventhal 1980, Mikula 1980). Each of these distributive rules is regarded as substantively different and as granting a distinct claim for reward (Törnblom and Vermunt 2007).

The multi-principle justice approach assumes the existence of several justice principles and rules that people employ whenever they consider fairness of distribution in different situations, i.e., they focus on the match between the nature of the actually applied distribution rule and the rule that is considered just and should have been applied. These models utilize several principles, notably equity, equality, and need, to represent justice (Cohen 1986, Deutsch 1985, Greenberg and Cohen 1982, Kayser and Schwinger 1982, Lerner and Whitehead 1980, Leventhal 1980, Leventhal, Karuza, and Fry 1980, Leventhal and Michaels 1969, Meeker 1971, Mikula 1980, Mikula and Schwinger 1978).

Drawing on this approach, Sabbagh and colleagues (1994) define social justice judgments (SJJ) as evaluations regarding the relative weighting to be assigned to different distributive rules when different social resources are distributed in society. (Degree of importance is ranked from "should be considered very much" to "should not be considered at all.") These evaluations indicate the individual's desired goals, e.g., perceptions of fairness and self-interest (Leventhal 1980); they do not focus on the collective aspects of justice processes (Hegtvedt and Johnson 2000). Not only can there be multiple dimensions of evaluation or legitimate sources for allocation, but there can also be multiple arrays of outcomes including intangible goods, services, rights, privileges, obligations, and responsibilities to be distributed. Different types of outcomes may correspond to different distributive rules or equity norms, a problem that cannot be utilized in the traditional exchange formulation and is outside the scope of the status value theory. In any social system, the stratification system is multidimensional (cf. Kimberly 1970) and indicates the presence of differences between individuals or groups in terms of status differences and power inequalities in addition to income differentials.

The Facet of Distributive Justice Theory

In this book, we conceptualize distributive justice judgments as complex evaluations whereby several principles of justice and social resources are considered simultaneously. Our main aim is to unveil the structure of distributive justice judgments, using this facet approach as a formal framework for expressing the multidimensionality of distributive justice judgments (Borg and Lingoes 1987, Borg and Shye 1995). Distributive justice theories can be constructed with diverse levels of analysis, including distributive good, community of justice, preconditions of distribution, the structure of principles and rules, and their content. We attempt to map their content domains so as to distinguish between political theories of global justice. The usefulness of any one level of analysis does not exclude the significance of other levels of analysis. Rather, it is more reasonable to argue that the whole picture of distributive justice is much better understood with the five levels of analysis considered together. An extended framework for the analysis of the multiplicity of conceptions of distributive justice, such as: What is to be distributed? Among whom are advantages and disadvantages to be divided (community of justice)? What are the preconditions for (re)distribution? What is the structure of the principle of distribution? What is the content of the principle of distribution?

Thus, in approaching the issue of the units of analysis, first we must understand the structure of distributive justice. In Figure 3.1 distributive justice is depicted as having multiple layers. Each layer consists of a distributive justice domain, which, in turn, comprises policy issues in which different kinds of relationships exist among policy actors. The distributive justice domain can be defined as a substructure identified by specifying a substantively defined criterion to examine the affinities and contrasts of distributive justice judgments in an effort to resolve the ethical dilemmas faced by international disaster management community.

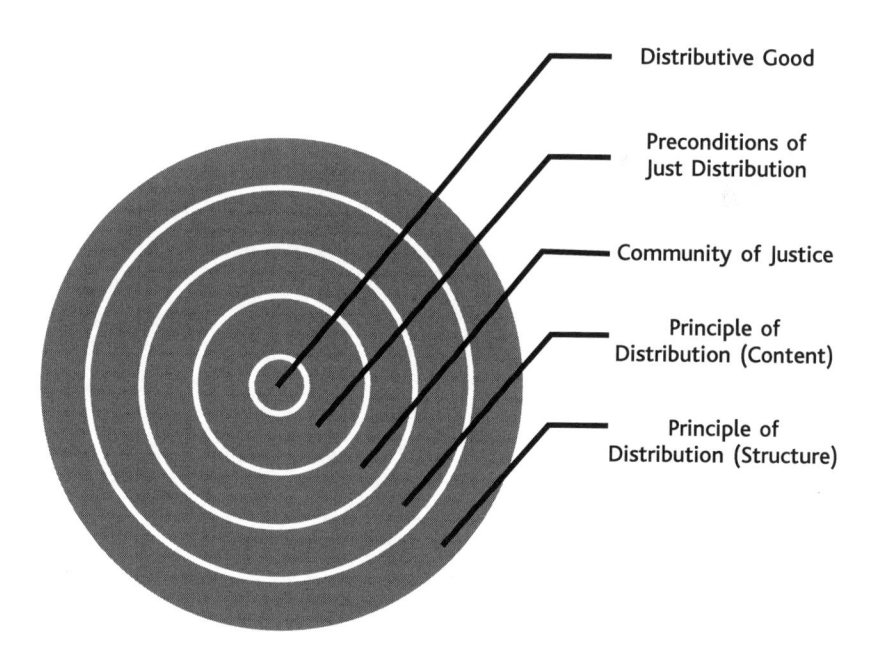

Figure 3.1. Multi-layers of Distributive Justice Evaluation.

Before considering different answers to these questions, it is essential to be aware of both the distinct kind of theoretical challenge that cosmopolitanism raises and the effects of distributive justice judgments. As noted, conventional theories of distributive justice tend to focus on benefits such as wealth and income. It is important, then, to ask whether this framework can effectively be extended to include the international disaster management community's burdens and benefits. A theory of distributive justice that is to be applied to international disaster management must, of necessity, address the question of whether the global dimensions of the issue make a morally relevant difference.

Cosmopolitanism and Global Distributive Justice

This section first questions the validity of adopting cosmopolitanism as a criterion for evaluating international disaster management practices for pressing distributive justice issues that operate within the jurisdiction of the state. Roots of cosmopolitan theory can be traced back to Stoicism, which viewed people as made of one universal spirit, living in brotherly love and mutual care. In the *Discourses*, Epictetus (1995, 20) maintained that "[e]ach human being is primarily a citizen of his own commonwealth; but he is also a member of the great city of gods and men, where of the city political is only a copy." In the Enlightenment era, cosmopolitanism was addressed by the philosophy of Immanuel Kant. Kant's cosmopolitanism refers to a global community where all have equal rights and access to goods. Thus, Kant's cooperative scheme that generates global relations of justice underlies citizens' participation in deliberative democratic processes by offering constructive criticism as they move collectively toward realization of perpetual peace and enlightenment.

Debate over the relative merits of cosmopolitan conceptions of distributive justice for collective action problems has been central to the writings of theorists such as Archibugi (2008), Beitz (1999, 2001), Brian Barry (1998), Caney (2005), Held (1995, 2003), Kok-Chor Tan (2000, 2004), Pogge (1989, 1994, 2002a, 2004), and Van Hooft (2009). Philosophies of global distributive justice are usually expounded in general terms relating to expansion of corresponding moral duties that can broaden the scope and responsibilities of justice in addressing the economic and social inequalities among societies. Cosmopolitans who discuss the role of institutions in establishing common principles and mutual actions do so from a global perspective. In defining the role of institutions in promoting social equity, they maintain that a reallocation of funds across state boundaries to equalize or compensate for existing inequality emerged from the global economic order (Pogge 2002b, Tan 2004).

The cosmopolitan emphasis on the structure of global economic institutions recognizes not only duties of charity (humanitarian and development aid) but much more substantial duties of justice in terms of equality. According to Tan, inequalities across nations are inherently "structural" inequalities that emerge "through the policies and method of operation of specific established institutional bodies" (Tan 2004, 26), such as the IMF or other international financial organizations. Therefore the logical conclusion for many is that such "patterned" injustices can be addressed only if the economic structure that causes such inequality is addressed. In other words, the circumstances of justice arise within an institutional structure, and lead to continuous and principled assessment of the structure of international political and economic institutions. Commentators question whether global institutions can be genuinely equitable, inclusive, and participatory (Brassett 2008, Castells 2003, Fraser 2005). This criticism is reinforced by David Held and Ronaldo Munck, who undertook the need to articulate the concerns over the equity of institutions at the global level such as in the UN (Held 2003, Munck 2005). Munck maintains that several UN initiatives focusing on global governance promote "market fundamentalism" alongside human rights. As these principles increase global risks, they create global inequities rather than addressing social exclusion. Munck questions the Western centrism underpinning the cosmopolitan approach: "Against all forms of racism and xenophobia, cosmopolitanism seems an attractive alternative. And yet the limits of these discourses are plain to see. They are imbued with the values of Western liberalism and therefore are constrained in their ability to offer a planetary solution" (Munck 2005, 117). However, while Munck supports Western centrism that lies behind the "'universalism' of human rights doctrines," he qualifies that "whether or not they are truly universal . . . human rights can be a powerful resource to combat oppression, even if they retain their character as a form of Western fundamentalism that would need to be deconstructed in any global process of progressive transformation" (Munck 2005, 162).

The debate over how global distributive devices can and should be used to promote an equitable distribution of resources reflects the cosmopolitan emphasis on the way institutional structure should be harmonized with understanding of the responsibility for unintended distributional consequences caused by the structure of associational embeddedness in a globalized world. For example, such relational global responsibility underlies the moral justification to compensate for the inequalities that are purportedly caused by economic integration and the current structure of international markets through redistributive transfers from affluent countries to the poor in developing countries (Pogge 2002a, 163). Given these arguments, it appears valid to evaluate the issues associated with international disaster management using formalized theories of cosmopolitan distributive justice. The following sections outline three of the most widely used theories of cosmopolitan justice that will be used to evaluate international disaster management practices.

Facet Analysis of Cosmopolitan Distributive Justice

In this section, we examine the nature of the relationship between Peter Singer's, Thomas Pogge's, and Amartya Sen's accounts of cosmopolitan distributive justice, which will allow us to capture the convergence and divergence, which simultaneously characterize this relationship.

What Is Distributed?

In his seminal 1972 article "Famine, Affluence, and Morality," Peter Singer uses an example of two children at risk: if for the sake of rescuing a child drowning in a nearby pond we are willing to ruin expensive new shoes, why are we unwilling to sacrifice a child dying overseas, which requires the same financial loss? Singer draws this argument on the following premise: "Suffering and death from lack of food, shelter, and medical care are bad" (Singer 1972, 231). According to Singer "food, shelter and medical care" are deemed necessary to alleviate suffering or death, because these goods are absolutely essential to increase the life spans and life chances of hundreds if not thousands of people (Singer 2009, 15–16; also see Singer 1972, 231).

By distribution of wealth, that is, the distribution of income, to aid poverty relief, individuals can fulfill their moral duty to help the poor for failures of global justice. In this ethical redistribution Singer generally addresses the international redistribution between rich and poor countries: "By not giving more than we do, people in rich countries are allowing those in poor countries to suffer from absolute poverty, with consequent malnutrition, ill health and death . . . If, then, allowing someone to die is not intrinsically different from killing someone, it would seem that we are all murderers" (2003, 162). Thus, for Singer, cosmopolitan distributive justice covers a fair share of available social means for the pursuit of an individual's life span, but whether people succeed or fail in their efforts is beyond the scope of distributive justice.

Thomas Pogge draws on the list of social primary goods proposed by Rawls (Pogge 1989, 147). Rawls's theory of social primary goods includes the following list (Rawls 1972, 92):[1]

1. The basic liberties (freedom of thought and liberty of conscience, etc.) are the background institutions necessary for the development and exercise of the capacity to decide upon and revise, and rationally to pursue, a conception of the good. Similarly, these liberties allow for the development and exercise of the sense of right and justice under political and social conditions that are free.

2. Freedom of movement and free choice of occupation against a background of diverse opportunities are required for the pursuit of final ends as well as to give effect to a decision to revise and change them, if one so desires.

3. Powers and prerogatives of offices of responsibility are needed to give scope to various self-governing and social capacities of the self.

4. Income and wealth, understood broadly as they must be, are all-purpose means (having an exchange value) for achieving directly or indirectly a wide range of ends, whatever they happen to be.

5. The social basis of self-respect are those aspects of basic institutions that are normally essential if citizens are to have a lively sense of their own worth as moral persons and to be able to realize their highest order interests and advance their ends with self-confidence.

According to Rawls, these goods are deemed necessary to the successful execution of any life plan, whatever that plan or conception of the good life, because these goods are absolutely essential for our self-determination and self-fulfillment as rational persons: "Things which it is supposed a rational man wants whatever else he wants. Regardless of what an individual's rational plans are in detail, it is assumed that there are various things which he would prefer more of rather than less" (Rawls 1993, 92). Rawls terms his primary goods theory a practical basis of interpersonal comparisons based on objective features of citizens' social circumstances open to view: "Provided due precautions are taken, we can, if need be, expand the list to include other goods, for example, leisure time, and even certain mental states such as freedom from physical pain" (Rawls 1993, 181–2).

However, Pogge extends the list of social primary goods to meet the global basic structure by ensuring that each participant enjoys secure access to a certain range of goods such as certain economic goods and political liberties and opportunities, including a right to security and a right to subsistence (Pogge 2002b, 48–9. Cf. Rawls 1993, 6–7). Pogge defends his broadening the scope of social primary goods to include "socioeconomic rights" on his list of human rights "as the relevant all-purpose means necessary for pursuing the interests and plans persons in modem democratic societies are in fact likely to have" (1989, 98).

He draws on the UN Declaration on Human Rights, which specifies, among other things, the right "to a standard of living adequate for the health and well-being of oneself and one's family, including food, clothing shelter and medical care" (UDHR 1948, Article 25).

Amartya Sen's theory offers to assess the kind of distributed goods in terms of human capabilities. According to Sen, a theory of primary goods must take into consideration "not only of the primary goods the persons respectively hold, but also of the relevant personal characteristics that govern the conversion of primary goods into the person's ability to promote her ends. What matters to people is that they are able to achieve actual functionings, that is the actual living that people manage to achieve" (Sen 1999, 74). The idea of functioning refers to anything a person could do or be, it "reflects the various things a person may value doing or being, varying from the basic (being adequately nourished) to the very complex (being able to take part in the life of the community)" (ibid., 75). If we want to make interpersonal comparisons of well-being we should use a category that encompasses both certain functionings and the capability to choose the attainment of such functionings,[2] that is, "the freedom to achieve actual livings that one can have a reason to value" (ibid., 73). Thus, Sen links capabilities not functioning with a liberal respect for individual freedom and also with an appropriate demand for individual responsibility: "In dealing with responsible adults, it is more appropriate to see the claims of individual on the society in terms of freedom to achieve rather than actual achievements" (Sen, 1992, 148).

The Community of Justice

In *Practical Ethics* (2003), Singer refers to the value of personhood to justify his extended view of the community of justice. According to Singer, the members of the community of justice share four main characteristics: (i) A rational and self-conscious being is aware of itself as an extended body existing over an extended period of time. (ii) It is a being capable of pursuing desires and decision making. (iii) It has a desire to continue living. (iv) It is an autonomous being. For Singer, if a being can display any of these characteristics then that being is a person and worthy of special consideration (2003, 87–100).

Based on his conceptualization of personhood, Singer rejects distance as a morally relevant criterion. For that, the scope of community of justice cannot be affected by location or national boundaries (Singer 2004, 148). According to Singer, "[t]he fact that a person is physically near to us, so that we have personal contact with him, may make it more likely that we shall assist him, but this does not show that we ought to help him rather than another who happens to be further away" (Singer 1972, 232).

Singer is aware of the fact that "[f]or many people, the circle of concern for others stops at the boundaries of their own nation—if it even extends that far" (Singer 2004, 152). Nevertheless, the application of new types of communications technologies and changes in transportation have been a driving force behind the greater visibility of distant suffering: "The complex set of developments we refer to as globalization should lead us to reconsider the moral significance we currently place on national boundaries" (Singer 2004, 171). Singer contends that "[f]rom the moral point of view, the development of the world into a 'global village' has made an important, though still unrecognized, difference to our moral situation." He notes that, in particular, "[e]xpert observers and supervisors, sent out by famine relief organizations or permanently stationed in famine-prone areas, can direct our aid to a refugee in Bengal almost as effectively as we could get it to someone in our own block" (Singer 1972, 232).

Indeed, most of the major problems faced by the global community such as poverty, overpopulation, and pollution affect almost every human being around the globe (Singer 1972, 233). Singer extends Benedict Anderson's conceptualization of a nation as an "imagined community," in which globalization "creates the material basis for a new ethic that will serve the interests of all those who live on this planet in a way that" can be used as a way to persuade individuals that they are actually members of a larger community, which embraces the entire world (Singer 2004, 12).

According to Pogge, the community of justice is the proper context and setting of distribution of primary goods. Such definition of community of justice manifests the Kantian idea of personhood that respects the equal worth and universal dignity of all persons. The members of the community of justice are all rational persons, capable of pursing a rational life plan based on their own perception of good and respect for others endowed with same capability "that enable them to be fully cooperating members of society over a complete life" (Rawls 1993, 183).

However, Pogge acknowledges that variation among members of the global community in the ability to convert access to resources (e.g., climate, environment, gender roles) is associated with "personal heterogeneities."[3] Personal heterogeneities are by no means natural, but result from past or present inequality in resource access under certain institutional arrangement. Thus, when given access to equal shares of resources some people will be better suited to meet their needs than others (Pogge 2004, 167–228).

Viewed in this way, Pogge suggests that "[r]esourcists believe that individual shares should be defined as bundles of goods or resources needed by human beings in general, without reference to the natural diversity among them. These goods might include certain rights and liberties, powers and prerogatives, income and wealth, as well as access to education, health care, employment, and public goods—with different lists and different weights specified by different resourcist views. Adherents of the capability approach hold, by contrast, that individual shares should be defined so as to take account of "personal characteristics that govern the conversion of primary goods into the person's ability to promote her ends."[4] Thus, an equalitarian capability criterion holds that, under a just institutional order, persons with mental or physical frailties or disabilities would receive more resources than others, enabling them to reach the same level of capabilities, the same level of opportunities to promote human ends, insofar as this is reasonably possible (Pogge 2002a, 192–3).

Pogge uses an example of diminishing women's suffering through institutional arrangements that secure equal treatment to both men and women, which in turn lead to challenge existing extra-institutional factors such as cultural attitudes and practices. According to Pogge: "Women's suffering in the world as it is does not result from social institutions being insufficiently sensitive to the special needs arising from their different natural constitution. Rather, it overwhelmingly results from institutional schemes and cultural practices being far too sensitive to their biological difference by making sex the basis for all kinds of social (legal and cultural) exclusions and disadvantages. Women and girls have a powerful justice claim to the removal of these barriers, to equal treatment (in a resourcist sense). If these barriers were removed, if our social institutions assured women of equal and equally effective civil and political rights, of equal opportunities, of equal pay for equal work, women could thrive fully even without any special breaks and considerations" (Pogge 2002a, 183).

While Pogge uses personal heterogeneities to denote the variation in ability to convert access to resources among community members, Sen uses personal heterogeneities to indicate the variation among individuals in ability to convert resources into functionings. According to Sen: "People have disparate physical characteristics connected with disability, illness, age or gender, and these make their needs diverse. For example, an ill person may need more income to fight her illness—income that a person without such an illness would not need . . . A disabled person may need some prosthesis, an older person more support and help, a pregnant woman more nutritional intake, and so on" (Sen 1999, 70). The reasons behind the differences in capability to function include: "(1) physical or mental heterogeneities among persons (related, for example, to disability, or proneness to illness); (2) variations in non-personal resources (such as the nature of public health care, or societal cohesion and the helpfulness of the community); (3) environmental diversities (such as climatic conditions, or varying threats from epidemic diseases or from local crime); or (4) different relative positions vis-à-vis others" (Sen 2005, 154). Sen links these opportunities to the perspective of freedom; it makes the idea of personal heterogeneities more inclusive and thus demands that the viewpoints of others, whether or not belonging to some group of which one is conceived as a member, receive appropriate attention. This opens the way to make assessments of people's state of existence in terms of people's actual doings and beings that people are able to choose. Sen contends that "[n]o theory of justice today can ignore the whole world except our own country" (Sen 2009, 173). Sen holds this demand for three main reasons: "Assessment of justice demands engagement with the 'eyes of mankind,' first, because we may variously identify with the others elsewhere and not just with our local community; second, because our choices and actions may affect the lives of others far as well as near; and third, because what they see from their respective perspectives of history and geography may help us to overcome our own parochialism" (ibid., 130).

Preconditions of Just Distribution

In Singer's example of the drowning child in the pond, we remain with no idea of the context and the circumstances surrounding the situation, such as: How did the child get into the pond? What are the consequences once the child is saved? According to Singer these questions have no bearing to the fact that we feel morally obliged to save the child. Thus, for Singer the only criteria that make one a candidate for moral consideration are the capacity of suffering and having preferences (2003, 57). As articulated by Singer: "The capacity for suffering [is] the vital characteristic that entitles a being to equal consideration. . . ." the capacity for suffering and enjoying things is a prerequisite for having interests at all, a condition that must be satisfied before we can speak of interests in any meaningful way . . . if a being suffers, there can be no moral justification for refusing to take that suffering into consideration . . . if a being is not capable of suffering, or of experiencing enjoyment or happiness, there is nothing to be taken into account. This is why the limit of sentience . . . is the only defensible boundary of concern for the interests of others" (2003, 57).

Preference is an additional criterion for evaluating actions as right or wrong. Preference is understood in a way that defines what the vulnerable (affected) individuals subjectively prefer: "For preference utilitarians, taking the life of a person will normally be worse than taking the life of some other being, since persons are highly future-oriented in their preferences. To kill a person is therefore, normally, to violate not just one, but a wide range of most central and significant preferences a being can have" (2003, 95). By drawing on the conditions of suffering and having preferences, distribution of aid considers the interests of everyone equally without giving any special weight to those with whom one has reciprocal relations.[5]

Pogge makes use of the original Rawlsian position by applying it to the global system in which the parties are bargaining as individuals for a just global structure. Rawls's theory of justice addresses the idea that individuals can hide behind a "veil of ignorance," where they are not aware of their social status, ethnicity, gender, nationality, or culture (Rawls 1972, 102). Behind that veil of ignorance, free from social constraints and self-interest, the individual can determine what justice consists of, especially just distribution of goods. Hence, the parties to the contract are rational and mutually disinterested (not all individuals wish to advance the interests of anyone other than themselves). They are deprived of the knowledge they would need to carry out their life plans according to their position in society, their skills, and the abilities they possess. Under the veil of ignorance, they all have the same general knowledge about human nature and the working of society. In their condition of ignorance they can only articulate general principles for distributing goods; that is, ensuring individual well-being.

The global veil of ignorance as envisaged by Pogge is required for a just global order that will be based on equality among the parties and fairness (Pogge 1989, 247, 272–3). According to Pogge, nationality should be considered as "just one further deep contingency (like genetic endowment, race, gender, and social class), one more potential basis of institutional inequalities that are inescapable and present from birth" (Pogge 1989, 247). Therefore, in order to downplay the arbitrary impact of nationality, Pogge suggests that the parties of the global original position would agree on a scheme of global justice "to be maximally supportive of basic rights and liberties, to foster equality of fair opportunity worldwide, and to generate social and economic inequalities only insofar as these optimize the socioeconomic position of the globally least advantaged persons" (Pogge 1989, 254). It enables individuals around the globe to come up with a global agreement on a list of human rights, which, over time, becomes more inclusive by developing a system of global economic constraints (Pogge 1989, 272). He continues by claiming that citizens of rich countries, to the extent that they do not challenge certain governmental policies, are violating the human rights of citizens of poor countries.

Sen proposes to replace the device of the global veil of ignorance with Adam Smith's idea of the impartial spectator in the *Theory of Moral Sentiments*. Smith uses the capacity of feeling natural moral sentiments such as sympathy to approve or disapprove our own and others' actions. Smith describes that as "[y]et to accurately feel what a third party is feeling in a given situation we must have information about how the situation came about, . . . our sympathy depends on our ability to approve of another's behavior because it is *appropriate* to the situation. Until we understand the context of another's behavior, we cannot know whether our own emotional response will be one of positive sympathy or negative revulsion, of approval or disgust." (Heilbroner 1987, 58).

Thus the ability to feel sympathy assists us to evaluate action as a third uninvolved party would evaluate it rather than on personal interests. Although the device of the veil of ignorance is effective in diminishing the effects of personal interests in distributive decision making, it is still unable to include in the deliberation in the original position the opinions of individuals belonging to different social groups. For Sen, the device of the impartial spectator helps to deliberate from the perspective of a "spectator" at a distance (Sen 2009). Sen also rejects the criterion of preference as a precondition for distributive justice. Sen draws on the notion of adaptive preferences to denote the circumstances of economic constraints and socially repressive traditions that influence the person's preferences or perceptions of satisfaction with her state of being as she tends to adapt her desires to what is viable. Thus, preferences depend on social arrangements; for example on gender biases such as making female secondary to male have influence on the intra-family distribution of food and health care (Sen 1999, 126). However, in a society where the conditions under distributive justice ensure basic capabilities to substantive freedoms, rather than preferences shaped by an acceptance of a given order or unjust background conditions, these freedoms "enhance the ability of people to help themselves and to influence the world" (Sen 1999, 18).

The Principle of Distribution (Content)

In his article "Famine, Affluence, and Morality," Peter Singer advocates the position that people in the developed world are morally obligated to provide assistance to poor people in the developing world (Singer 1972). Two principles underlie such position. The first principle concerns the need to reduce suffering since "that suffering and death from lack of food, shelter, and medical care are bad" (Singer 1972, 231). The second principle refers to how much we should be obligated to sacrifice in order to prevent suffering: "If it is in our power to prevent something bad from happening, without thereby sacrificing anything of comparable moral importance, we ought morally to do it" (Singer 1972, 231). "Comparable moral importance" is meant by Singer as an amount or set of resources that we could probably sacrifice significantly without reducing ourselves to certain level of poverty. As articulated by Singer: "Just how much we will think ourselves obligated to give will depend upon what we consider to be of comparable moral significance to the poverty we could prevent: stylish clothes, expensive dinners, a sophisticated stereo system, overseas holidays, a (second?) car, a larger house, private schools for our children, and so on" (1972, 231–2). He claims that none of these goods is likely to be of comparable significance to the reduction of poverty. When viewed from an applied ethics perspective, Singer argues that 10 percent of our total income should be devoted to prevent poverty: "Any figure will be arbitrary, but there may be something to be said for a round percentage of one's income, say 10 percent—more than a token donation, yet not so high as to be beyond all but saints. . . . Some families, of course, will find 10 percent a considerable strain on their finances. Others may be able to give more without difficulty. No figure should be advocated as a rigid minimum or maximum; but it seems safe to advocate that those earning average or above average incomes in affluent societies, unless they have an unusually large number of dependents or other special needs, ought to give a tenth of their income to reducing absolute poverty. By any reasonable standards this is the minimum we ought to do, and we do wrong if we do less" (Singer 2003, 246).

Pogge's human-rights-based framework applies the two principles of distributive justice suggested by Rawls at the global level. The first principle, also known as the equality principle, addresses equality of liberties, which Rawls construes in a negative sense as freedom from socially imposed "constraints" or "interferences" (Rawls 1972, 202). Thus, according to Pogge the basic structure of the global system should be governed by the principle that involves "a just and stable institutional scheme preserving a distribution of basic rights, opportunities and . . . goods that is fair both globally and within each nation" (Pogge 1989, 256). The Universal Declaration of Human Rights (UDHR), for example, can be seen as a reflection of the globalized first principle of justice endorsed by Pogge. The Universal Declaration of Human Rights sets forth the principle according to which "[e]veryone has the right to a standard of living adequate for the health and well-being of himself and of his family, including food, clothing, housing and medical care. Everyone is entitled to a social and international order in which the rights and freedoms set forth in this Declaration can be fully realized" (UDHR 1948, Articles 25 and 28).

Accordingly, the second principle, the difference principle, must apply globally to individuals and not to states. Rawls's second principle allows for unequal distributions of some primary goods as long as these inequalities produce benefits to the least well-off members of society and they are attached to positions and offices open to all. The former requirement is called the "difference principle," and this latter requirement embodies a principle of fair equality of opportunity (Rawls 1972, 74). It follows that the extent of the inequalities justified by the difference principle depends on certain contingent psychological facts about those people whose skills and talents are such that their greater exercise will improve the position of the worst-off. The difference principle permits some differences only to the extent that they are to everyone's mutual advantage, while the principles of fair equality of opportunity and equality of liberty set some limits on these differences. By extending Rawls's second principle to the global level, Pogge suggests that societies should have a fair opportunity for economic development, where the criteria of fair equality of opportunities should be set under the conditions of the global veil of ignorance. Thus, permitting global inequality is unfair to those who are in the position of the worst-off. For that, Pogge proposes to alter the unjust global institutions since the conditions of life all over the world are deeply affected by the global institutions. This leads him to come up with a global redistributive scheme to compensate the poor for their exclusion from sufficient access to natural resources that is realized in the Global Resource Dividend (GRD) (Pogge 2002a, 196–215). Under the scheme of a GRD, the global poor have an "inalienable stake in all scarce resources" since national borders have no moral value, and are created over time through power relations based on coercion and violence (Pogge 1994). The GDR will force nations to place a dividend (tax) on any resources they use or sell that will result as a "tax on consumption" to alleviate poverty (Pogge 1998).

In his book entitled *The Idea of Justice*, Sen criticizes principles of justice that derive from the Rawlsian veil of ignorance. In his criticism, Sen actually exposes the principles that should govern the just distribution: "First, it concentrates its attention on what it identifies as perfect justice, rather than comparisons of justice and injustice.... Second, in searching for perfection, transcendental institutionalism concentrates primarily on getting institutions right, and is not focused directly on the actual societies that would actually emerge" (Sen 2009, 5–6). Sen admits only that "[t]he interpersonal comparisons that must form a crucial part of the informational basis of justice cannot be provided by comparisons of holdings of means to freedom (such as 'primary goods,' 'resources,' or 'incomes')," but rather must be measured using capabilities (Sen 1990, 112).

He insists that a "capability-centered view gives us a better understanding of what is involved in the challenge of poverty." In addition, poverty in the modern sense is not just low income, but "the lack of substantive freedom" (Sen 1992, 151). Thus, Sen urges that the principle of distributive justice should allow everyone in the society to enjoy freedom. The rationale of sufficientarianism behind Sen's principle of justice requires basic capability equality—policies and actions should be pursued so that each person equally has freedom to attain a sufficient level of basic capabilities. However, Sen is less explicit about the extent of freedom a society should promise to its members such as whether it is the duty of the society to assist people to reach every goal they value by economic rearrangements. Sen does not explain exactly how to rank the order of different functionings and hence different capabilities to achieve functionings as the measurement of well-being, but remarks that "economic unfreedom, in the form of extreme poverty, can make a person a helpless prey in the violation of other kinds of freedom" (Sen 1999, 8).

The Principle of Distribution (Structure)

Singer's principles of justice are founded on the structure laid by universalizability which "requires us to go beyond 'I' and 'you' to the universal law, the universalisable judgment, the standpoint of the impartial spectator or ideal observer, or whatever we choose to call it" (2003, 12). The way to justify this structure, according to Singer, uses the Principle of Equal Consideration of Interests. The Principle of Equal Consideration of Interests considers that one should regard everyone's interests equally when making decisions. The fact that all beings are capable of suffering dictates equal consideration that all beings are equal (2003, 57). According to Singer: "This means that we weigh up interests, considered simply as interests and not my interests, or the interests of Australians, or of whites. This provides us with a basic principle of equality: the principle of equal consideration of interests" (2003, 19). The universe should function as an extended structure that cannot morally justify personal relationships: "The fact that a person is physically near to us, so that we may have personal contact with him, may make it more likely that we shall assist him, but this does not show that we ought to help him rather than another who happens to be further away. If we accept any principle of impartiality, universalizability, equality, or whatever, we cannot discriminate against someone merely because he is far away from us (or we are far away from him)" (Singer 1972, 232).

Pogge stems his prescribed global principles of justice from a position of impartiality by imagining the global institutional arrangements we would design if we ignore the most identifying facts about ourselves such as historical, social, cultural, or personal characteristics, in order to adopt a position that goes beyond the differences between individuals and societies (Pogge 1989, 247). Pogge is concerned with those structures that connect institutions with individuals to regulate international economic interaction: "All human beings are now participants in a single, global institutional order—involving such institutions as the territorial state, a system of international law and diplomacy, as well as global economic system of property rights and markets for capital, goods and services" (Pogge 2002a, 168). Thus, transnational social structures, which manage the multiple transactions of an interconnected world, must account for the asymmetry in the global sphere inasmuch as some people become more vulnerable to coercion, domination, and deprivation due to "an elaborate system of treaties and conventions about trade, investment, loans, patents, copy rights, trademarks, double taxation, labour standards, environmental protection, use of seabed resources" (Pogge 2004, 263). Viewed in this way, any structural condition of asymmetric concern is accepted to be fair only on the basis of a symmetric concern and respect for all individuals.

Sen advocates the universal grounding offered by both Singer and Pogge to consider instances of injustice. However, he offers to broaden the universal impartial evaluative basis to include an "open impartiality," which according to Sen, is "the procedure of making impartial assessments" that invokes judgments both from within and from without any given community, seeking to identify the "disinterested judgments of 'any fair and impartial spectator'" (Sen 2009, 123). Sen's notion of "open impartiality" incorporates two aspects to the meaning of "open." One aspect is associated with the need to introduce distance from any particular position to ensure being "open" to considerations from anyone; the second aspect concerns the epistemological conditions that are required of disinterested judgment about justice and its demands by a subject (evaluator) (Sen 2009, 128). This structure allows taking into account all the arguments and counterarguments by the "eyes of mankind" (Sen 2009, 130) as "positional objectivity" (Sen 1993, 2002). Sen's conception of capability explicitly focuses on "positions" that are not easily exchangeable relative to alternative social rules. Rather, each individual in as much as she will be an "object" (recipient) of social rules is expected to provide information as "subject" (evaluator) of social rules about such particularities, being a witness or an expert of those characteristics or experiences that cause different kinds of disadvantages.

Summing Up

The purpose of this chapter was twofold: first to establish the criteria of distributive justice that could be applied to equity issues in international disaster management, then to present three theories of global distributive justice that aim to produce different principles for equitable distribution and access to disaster response and relief resources and services. The global just philosophies of Singer, Pogge, and Sen were chosen as benchmarks for this analysis because they represent different schools of thought on global justice while sharing the demand that a fair share of resources is necessary, as is access to opportunities of welfare as a component in achieving the global objectives of distributive justice at the global level.

The important differences between the doctrines lie in the conditions relevant for determining whether the global distribution of goods is fair. Singer upholds goods that are necessary to alleviate suffering or death; therefore the measure of a person's condition, which determines whether their conditions are equal or unequal, are construed as the elimination of suffering or as preference satisfaction. Pogge proposes to extend the Rawlsian social primary goods approach to meet the demands of the global scheme as general-purpose means that will be useful for promoting a wide variety of rational plans of life. Sen suggests the measures of functionings and functioning capabilities by addressing inadequacies in the Rawlsian social primary goods extended by Pogge and the preference utilitarian pursued by Singer as a basis for interpersonal comparison measures.

Consequently, the principles of global distributive justice that derive from the justification of the conditions relevant for determining whether the global distribution of goods is fair, vary among these philosophers. While Singer's theory of justice is concerned with the duty to alleviate suffering when this can be done, even when there is a significant cost to ourselves, Pogge draws on negative duties, that is, not to violate the human rights of others (namely, citizens of poor countries) or support institutions that do so. Sen rejects the conceptualization of ends of well-being provided by both Singer and Pogge, replacing them with the concept of people's capabilities to function. According to Sen the principle of justice must ensure that each person has the capability to gain a sufficient level of attainment to each and every one of the functionings, that is, their effective opportunities to undertake the actions and activities that they want to engage in, and be whom they want to be.

4

The Dependency Syndrome

Disasters are first and foremost a major threat to development, and specifically to the development of the poorest and most marginalized people in the world. Disasters seek out the poor and ensure they stay poor.

—Didier Cherpitel, Secretary General of the International Federation of Red Cross and Red Crescent Societies (IFRC), 2001

The idea of dependency syndrome in disaster emergency contexts often denotes concerns about the possible negative impacts of humanitarian aid or emergency relief. Although international relief aid used in humanitarian emergencies aims at preserving human lives and livelihood assets, it may lead to unintended, adverse consequences associated with a sense of undermining local capacity and responsibility of government agencies to deliver aid services and resources to meet basic humanitarian needs. It is then suggested that concerns about dependency are more about demarcating the boundary between international relief agencies' responsibility and governments' responsibility to provide relief and response assistance. Thus, this book focuses more on how the conceptualization of dependency influences what different actors engaged in disaster management, such as international aid agencies and governments, do in terms of practical ethics (global distributive justice) rather than on whether vulnerable communities are or are not dependent.

This chapter aims to provide some clarity about the term *dependency syndrome* when used within the field of international disaster management. The meanings attached to this concept in the field of international disaster management carry important consequences on humanitarian aid providers to justify a withdrawal from response and relief efforts, or on governments to claim overall control on the whole response and relief process and means. Thus, the discourse on dependency syndrome within the field of international disaster management will draw on existing definitions of dependency in the literature of humanitarian aid and how it can be usefully applied in the context of international disaster management in providing justifications for action or inaction.

What Is Dependency?

The concept of dependency entangles various meanings when used in different contexts. When dependency is used in the context of humanitarian aid relief it is often associated with negative values that undermine individuals' and communities' self-reliance and sustainability (Harvey and Lind 2005). But before tracing the way in which dependency confounds short-term and longer-term negative effects of international aid intervention, let us introduce the existing positive aspects of dependency in the literature.

In a general way, dependency is one of the constitutive features for the existence of any human community. As stated by Dean (2004, 194): "Interdependency is an essential feature of the human life course and the human condition. One might argue that it is constitutive of our humanity and the achievement of human identity. This is neither new nor radical. It is captured for example, in the timeless aphorism attributed to the Xhosa people of South Africa—'A person is a person through other persons.' Personhood is founded in and through dependency on other persons." As such, dependency is necessary to shape a system of rights and responsibilities based on the rationale of caring for others' needs.

According to Lensink and White, who provide a somewhat positive definition to aid dependence (1999): "A country is aid dependent if it will not achieve objective X in the absence of aid for the foreseeable future." The definition proposed by Lensink and White addresses the idea that external aid should not necessarily lead to "wrongdoings" and that countries that rely on high levels of aid regardless of whether they make progress toward some development objective should not be regarded as "dependent" on the aid for achieving their development objectives. At the same time, Lensink and White come up with a short list of "aid dependent," namely developing countries where aid remains an ineffective mechanism for meeting development objectives.

Lentz and Barrett (2005) suggest defining dependency in a positive manner when viewed from a welfarist approach. Dependency is regarded as a better alternative to welfare enhancement rather than an alternative coping mechanism held by individuals or communities that cannot meet their immediate basic needs without reliance on external assistance. Thus, one can think of dependency as similar to a provision of subsidy. As such, aid can contribute to economic development in order to support local efforts and competencies: revenue collection, investment in physical and human capital, and institutionalization of developmental state[1] (Lappe and Collins 1977). Depending on external assistance may be regarded as a coping mechanism by active rather than passive recipients of aid that enables people to maintain their livelihoods and prevents a slide into destitution (Ellis 2000).

The debate on the undesirable effects of dependency arose during the 1960s and 1970s left-wing critiques of Western aid to the developing world. Thus, when dependency became associated with Marxist economic development theories it was framed as the antithesis of development, resulting in unequal power relations between rich developed countries and poor developing ones. The fact that developing countries are seen as dependent on continuing humanitarian aid assistance (e.g., food aid) may undermine local development programs that rely on individuals' and communities' competencies and contribution to labor: "Adjustments to high levels of aid over a number of years are all part of what is often referred to as 'aid dependence.' A dependence on high aid inflows is not necessarily a bad thing. If the aid is used productively to promote social and economic progress, its net effect is likely to be highly positive for development in the country receiving the aid. But where the aid is ineffective, it is important to consider the potential negative effects of that aid" (Lancaster 1999, 494). Following Roger Riddell and Rehman Sobhan, the negative effects of aid dependence should be considered within "that process by which the continued provision of aid appears to be making no significant contribution to the achievement of self-sustaining development" (1996, 24). In other words, humanitarian aid may challenge initiative and capacity building efforts at both individual and community levels.

During the 1970s and 1980s, development economists and policy-makers such as Paul Streeten, France Stewart, and Mahbub ul Haq issued an international development program following a basic need approach (BNA) (Gasper 1999, Sen 2000, Streeten et al. 1981). The BNA approach aimed at providing a positive aspect of development by drawing on a broader sense of human well-being: "The objective of a basic needs approach to development is to provide opportunities for the full physical, mental, and social development of the individual" (Streeten 1979, 136). The BNA suggests that a society cannot be defined as developed unless it offers the opportunity for all its citizens to meet their basic needs. Thus, the BNA interprets basic needs as opportunities of valuable functionings: "In addition to the concrete specification of human needs in contrast to abstract concepts, and the emphasis on ends in contrast to means, the basic needs approach encompasses 'nonmaterial' needs. They include the need for self-determination, self-reliance, political freedom and security, participation in decision making, national and cultural identity, and a sense of purpose in life and work" (Streeten 1979, 136).

When dependency is situated within a wide literature around development, it denotes certain behaviors generally seen as negative and to be avoided—improvidence, laziness, and fatalism: "Dependence is decadent—its byproducts are laziness and degeneration, poverty and crime. It is a way of life transmitted to the young, generating cycles of dependency. Dependency justifies, even compels, negative judgments. As a result, both political parties enthusiastically endorse punitive measures to spur dependents towards independence" (Fineman 2001). Thus, the unintended consequences of dependency are reflected in the categorization of individuals and groups as dependent and vulnerable, which often becomes highly politicized, leading to stigmatization of those groups (e.g., single mothers, the elderly, ethnic minorities). In addition, the large amount of relief cash donations may create incentives for corruption.

However, framing these concerns in terms of dependency runs the risk of furthering other and more negative forms of dependence such as justifying cutting back humanitarian aid for vulnerable people. In situations where disaster overwhelms local capacities, cutting back relief resources and services will limit access to basic welfare and increase exposure to violence.

Thus, the focus of the discourse around dependency in international disaster management should not be on avoiding dependence in emergency relief, but rather on providing equitable reliable and transparent assistance. International aid agencies need to be concerned about the normative effects of their aid. The debate then shifts to the responsibility of aid agencies to ensure that people at risk are able to reliably depend on receiving assistance. The global distributive justice framework is suggested to address some of the negative consequences of dependency in disaster response and relief efforts.

Analyzing Dependency in International Disaster Management

What Is Distributed?

As seen, dependency is assessed by the contribution that aid makes to people's livelihoods. In this sense, dependence on response and relief efforts is a mechanism to meet subsistence needs in times of large-scale disaster. Response and relief efforts are regarded as the most visible disaster management function at the international level (Coppola 2011). This phase includes those activities that directly address immediate needs of affected people, namely medical care, water and sanitation, food, and shelter. The international response to the tsunami of December 2004 organized these basic goods in nine clusters at a global level led by aid agencies (Tsunami Evaluation Coalition 2006, 26):

- Camp coordination and management—UNHCR (for conflict-generated IDPs) and IOM (for natural disasters)
- Emergency telecommunications—OCHA as overall process owner; UNICEF for data collection; WFP for common security telecommunications service
- Early recovery (formerly called reintegration and recovery)—UNDP
- Emergency shelter—UNHCR (for conflict-generated IDPs) and IFRC (for natural disasters)
- Health—WHO
- Logistics—WFP
- Nutrition—UNICEF
- Protection—UNHCR (for conflict-generated IDPs) and UNICEF and OHCHR (for natural disasters)
- Water and sanitation—UNICEF
- Food—WEP

For example, at the outbreak of Typhoon Haiyan (also known as Yolanda) in the Philippines, the Philippines government initiated a list of needed resources and services through the United Nations Office for the Coordination of Humanitarian Affairs (UNOCHA). The list included food, water, sanitation and hygiene kits, shelter, medicine, debris clearing, and logistics hubs to support the sustainable delivery of aid (UNOCHA 2013).

The list is applicable to communities and individuals who are no longer able to meet their subsistence needs because of disaster. As seen, Singer considers these goods (e.g., food, shelter, and medical care) as necessary to alleviate suffering or death. Pogge's theory of distributed goods adds the right to security to the right of subsistence. The right to security applies during the disaster response period in which the social order of the affected population is completely disrupted. The impact of insecurity on disaster response and relief efforts is also marked by rising casualty incidences among international aid workers engaged in response and relief efforts (Stoddard, Harmer, and Haver 2006, Tennant, Doyle, and Mazou 2010). The Montreu X Humanitarian Retreat section on the theme of "Safety and Security in Humanitarian Action" was concerned with issues of how to support good practice and enhance operational security for humanitarian action. The review addresses the need to assess the distribution of funding for security in international disaster management and individual agency security management[2] (Stoddard and Harmer 2010).

Sen calls for better analysis of distributed goods of people's own capacities, which aim to better reflect local capacities and internal competencies and dispositions that can be cultivated and facilitated by external aid and resources, but never provided by them (Sen 1985). Viewed in this way, response and relief efforts should be distributed to the extent that does not undermine the affected community's autonomy and freedom to choose and to act as "greater freedom enhances the ability of people to help themselves and to influence the world" (Sen 1999, 18). Dependence as a concept used by international disaster management needs to describe affected communities and individuals that are impeded from acquiring basic capabilities and meet their subsistence needs because of disasters. Basic needs operationally include food, water, shelter, and hospital beds during a disaster event. Sen is correctly concerned that these goods will be viewed as only "minimal needs," that only physical needs are what count rather than as opportunities for other valuable functionings. Focusing on "basic needs and nothing more" lends itself to down-sizing well-being and moral responsibilities (Sen 1984, 515).

Community of Justice

In extreme events that often overwhelm the ability of people to meet their subsistence needs, international aid agencies must attempt to assess the vulnerability of affected individuals, groups, communities, and countries in order to effectively address the victims' needs with their response services and resources. Dependency can be determined by establishing vulnerability criteria of relief recipients. Singer and Pogge define the boundaries of the community of justice by drawing the criteria of rationality, capacity of suffering, and recognition of preferences and desires. Sen also admits that reasoning within a social sphere is a defining feature of members of the community of justice. However, Sen claims that economically and socially repressive constraints influence people to adapt their preferences to what is feasible. By drawing on functionings, Sen provides a more objective assessment of members of the community of justice in terms of people's actual doings and beings (Sen 2000, 24–25). Within the context of dependency of relief assistance, certain groups in society such as women, children, and the elderly may face different types of vulnerabilities and so may be dependent in different ways on external relief assistance. However, following Pogge's criticism of the extent to which the capabilities approach may lead to stigmatizing those with fewer, or less valued, capabilities, targeting groups as vulnerable in disaster events by international aid agencies might reinforce stereotypes and relations of dependency. The stereotype of relief recipients as helpless is suggested to inform their perceptions concerning the role that they are expected to play to gain the approval of the help providers and succeed in receiving aid as a strategy to survive in the competition over scarce humanitarian aid resources. Thus, international disaster relief agencies need to gain meaningful knowledge of local communities and assist affected communities to articulate their needs in a way that enables relief agencies to act on them (Ireni Saban 2014). If not done, evaluations of vulnerable people may not meet objective need assessment measures of services precisely because they do not include the interpersonal aspect of service delivery so critical to the end user. It is then possible to move away from the negative aspects of dependency by assuming a positive role of international disaster relief agencies to treat local communities as more active, rather than passive, recipients of assistance, and to provide opportunities to change social relations. Sen stresses the need for reorientation of the institutional approach from one of providing goods and services to passive recipients to a BNA one that provides countries with genuine opportunities to dictate their development track: "The ends and means of development call for placing the perspective of freedom at the center of the stage. The people have to be seen, in this perspective, as being actively involved—given the opportunity—in shaping their own destiny, and not just as passive recipients of the fruits of cunning development programs" (Sen 1999, 53). Thus, following the BNA approach, Sen insists on distribution of actual freedoms rather than commodities: "The actual freedoms enjoyed by different persons—persons with possibly divergent objectives—to lead different lives that they can have reason to value" (Sen 1990, 114).

Preconditions of Just Distribution

Singer establishes our obligation to act morally dependent on the positive right to distribute aid efforts to support affected communities. However, Pogge justifies our obligation to be morally dependent in terms of negative duties to alleviate suffering as we are guilty of imposing and benefiting from global poverty (Pogge 2007, 219). Pogge's view of global inequalities echoes colonialism, which heightened dependence on an inherently exploitive relationship between states. The political roots of dependency then undermined the ability of poor citizens to become self-sufficient. These conditions, according to Pogge, reward dependency syndrome, which is continuing reliance on external assistance. Due to these contextual conditions, Pogge sketches a causal link between the operations of the global market and arising inequalities (or harms to the poor) at the global level: "Dominant Western countries are designing and upholding global institutional arrangements, geared to their domestic elites, that foreseeably and avoidably produce massive deprivations in most of the much poorer regions of Asia, Africa, and Latin America The underling casual claim regarding the conditions of justice" (Pogge 2008, 25). The underlying causal argument of justice justifies an extended sense of duty to incorporate positive as well as negative duties not to harm that sets two requirements for justice: foreseeability and avoidability. Thus, compensation in contrast to assistance may be "offsetting an unjust institutional redistribution from the poor to the rich" (Pogge 2004, 278) under the conditions where harm is foreseeable (e.g., giving international institutions incentives to corrupt poor countries' officials) and whether they were reasonably avoidable (Pogge 2008, 26). Sen also supports the view that there are economic and social conditions that may not encourage people's critical assessment or ability to question their preferences and interests

(Sen 2000, 24). Drawing on the preconditions of justice, all three philosophers might agree at some point that dependency on international relief agencies leads to tension between governments and relief agencies and the desire of relief agencies to gain control over the delivery of their resources. For that, it is necessary to promote greater levels of local transparency and accountability to beneficiaries so that aid recipients know what they are entitled to and how likely it is to be provided. A recent example of the likelihood of disaster relief assistance embedded in exploitive structures can be found in the case of the Haiti earthquake (2010). On January 12, 2010, Haiti faced a 7.0 magnitude earthquake, which caused the deaths of 100,000 to 160,000 and massive damage to infrastructures such as hospitals, air, sea, and land transportation, and communication systems. At the outbreak of the earthquake, due to widespread devastation and damage throughout Port-au-Prince and elsewhere to infrastructure, food supply lines were down. Foreign food aid agencies immediately mobilized provisional needs for the earthquake-affected population. Between January and June 2010, USAID sent 214,000 metric tons of food aid to Haiti (USAID 2010, 15). Irregularity in the distribution of food aid led to price fluctuations of basic commodities especially rice, hindering the ability of mini-wholesalers at the local level to buy rice to sell to consumers. The high amount of food aid depressed commodity prices and resulted in lower income of Haitian corn producers and farmers. Foreign food aid forced Haitian entrepreneurs out of market competition (Webster 2012). The main concern therefore is not dependency per se, but the way in which dependency is structured, namely by causing disincentives for local producers and farmers.

The influx of donated rice following the Haiti earthquake demonstrated Pogge's argument of unequal power relations at the global level. It is then important to consider the way Haiti became dependent on continuing food aid, which was tied up with unequal trade relations between Haiti and the United States. During the 1980s, Haiti was a self-sufficient rice producer. However, President Clinton, who was then nominated as UN Special Envoy to Haiti, developed a close relationship with Haitian President Jean Bertrand Aristide. This relationship paved the way for American rice exporters' access into the Haitian local market (Webster 2012). By 1994, two years after Clinton's nomination, American Rice Inc. was already netting US$ 373 million in rice sales annually (Chavla 2010, Kenny 2011). In 2010, Clinton acknowledged the negative implication of his policy, which resulted in a reduction of rice consumed in Haiti from 47 to 15 percent of the rice supply: "Since 1981, the United States has followed a policy, until the last year or so when we started rethinking it, that we rich countries that produce a lot of food should sell it to poor countries and relieve them of the burden of producing their own food, so, thank goodness, they can leap directly into the industrial era. It has not worked. It may have been good for some of my farmers in Arkansas, but it has not worked. It was a mistake. It was a mistake that I was a party to. I am not pointing the finger at anybody. I did that. I have to live every day with the consequences of the lost capacity to produce a rice crop in Haiti to feed those people, because of what I did. Nobody else."[3]

American industry has actually benefited from the influx of donated rice in the aftermath of the disaster: "Haiti is even more awash in rice while American agribusiness makes billions of dollars every year through generous government subsidies" (Webster 2012). Paul Farmer argued that the interests of local Haitian producers and farmers were excluded in food aid provision decision making: "If anyone had real cause for complaint, it was—and still is—the Haitian people themselves, so long excluded from any meaningful discussion of their fate. To a list of grievances spanning at least two centuries, they added the inability of state and non-state providers to ensure basic succor to those in great need, in spite of the large presence of humanitarians and NGOs" (Farmer 2011, 43–44).

THE PRINCIPLE OF DISTRIBUTION (CONTENT)

Singer's principle of justice includes the mutual assistance principle, according to which people in affluent countries are morally obligated to help those who are unable to meet their subsistence need, if "without thereby sacrificing anything of comparable moral importance, we ought morally to do it" (Singer 1972, 231). Pogge suggests a way to apply the Rawlsian difference principle at a global level. He suggests conceiving the terms of international cooperation so that the social inequalities due to natural contingencies (the distribution of natural assets) are inclined to optimize the worst representative individual share (Pogge 2002). However, Pogge's view of owing a degree of assistance to compensate for the harm caused by the exploitive structure at the global level seems inconsistent when applied to natural disasters such as the case of Hurricane Katrina. Contemporary studies of the response and relief efforts following 2005 Hurricane Katrina have stressed that Hurricane Katrina exposed some of the major breakdowns of governance performance. Failures of the federal government's response to the needs and concerns of political constituents and public stakeholders were promulgated by lack of collaboration across the public, volunteer, and private sectors. The case of Hurricane Katrina then raises questions about Pogge's dismissal of positive duties in times of catastrophic disasters when ruling institutions failed to provide appropriate preparedness and mitigation planning and therefore hold responsibility for alleviating suffering.

In an interview with Thomas Pogge in 2007, he was asked: "Is your emphasis on negative duties, i.e., the duty not to harm, perhaps a pragmatic one, rising from the belief that most severe poverty is the direct result of rules imposed by rich countries? If we succeeded in changing the economic structure of the world such that rich countries no longer harmed poor countries, would we no longer have a duty to aid each other? Or in cases where harm is not a result of international policies, do we have a duty to aid? For example, when New Orleans was struck by a natural disaster, various charities, including the hungersite.com, made international appeals for help. Do you feel that their appeal was in any way legitimate, or rather, that citizens of the world had no duty to respond?"

Pogge replied: "Yes, this emphasis is in part pragmatic. I do believe that most severe suffering in this world would be avoided if the rich countries merely fulfilled their duty not to harm. I also find, especially in the Anglophone countries, a great reluctance to take positive duties seriously . . . Your New Orleans example is flawed in an interesting way. Yes, New Orleans was hit by a disastrous storm (Katrina). But the city was flooded because the levies were insufficient and had been known to be so for a long time. The great harm people suffered was caused by a confluence of natural and social factors. And I would then have formulated the appeal—certainly to compatriots—differently. Not: 'Dear fellow citizens, please help us, we were hit by a storm.' But rather: 'Dear fellow citizens, due to a grotesquely unjust allocation of infrastructure spending by the federal and state governments, favoring white over black, affluent over poor, and Republican over Democrat neighborhoods, we have been exposed to a substantial risk of devastating flood. This flood has now come to pass, and the damage it does is your responsibility. You must now do what you can to minimize the harm you will have caused.'"

Sen's principle of justice requires basic capability equality—policies and actions should be pursued so that each person equally has freedom to attain a sufficient level of basic capabilities. Although Singer's altruist and Pogge's reciprocal principles of distributive justice at the global level can form an important social safety net to meet immediate needs arising from disaster, Sen's principle, which stresses some degrees of freedom and autonomy as well as needs of health, food, and shelter, has a more realistic chance of integrating disaster relief and development objectives by allowing local communities and NGOs to exercise their capability for deliberation and informal safety nets and support in the face of a disaster event.

THE PRINCIPLE OF DISTRIBUTION (STRUCTURE)

Pogge's just global institutional principles that shape our responsibilities and rights can conceivably be structured on interdependency: "There is an injustice in the economic scheme, which it would be wrong for more affluent participants to perpetuate. And that is so quite independently of whether we and the starving are united by a communal bond" (Pogge 2008, 182). Singer also argues for impartiality and universal structure behind the obligation of individuals who can provide assistance to do so regardless of any causal relationship with poverty. Sen advocates the relevance of a more plural grounding for applying global justice principles beyond the relations between sovereign countries. Thus, local parochialism ensures open impartiality as deliberations of justice at the community level should be represented and recognized: ". . . assessment of justice demands engagement with the 'eyes of mankind,' first, because we may variously identify with the others elsewhere and not just with our local community; second, because our choices and actions may affect the lives of others far as well as near; and third, because what they see from their respective perspectives of history and geography may help us to overcome our own parochialism" (Sen 2009, 130). We argue that all philosophers share the concept of interdependency as a basis for recognizing our duties to provide assistance to distant others at the global community level. This structure establishes cooperative forms of responsibility and solidarity. What is being suggested by Sen is to enlighten the parochial views of each society in global deliberation of relief and aid allocation so that relief will be more participatory and accountable.

Summing Up

This chapter has attempted to apply global distributive justice approaches to analyze the positive and negative aspects of dependency within the context of international disaster management. The motivation behind the use of global distributive justice theories is to challenge the common tendency of dependency discourse to put the blame on external disaster relief aid rather than on poverty and existing exploitive economic and social structures that create vulnerability. Singer, Pogge, and Sen agree that international humanitarian aid agencies should distribute relief assistance to meet immediate subsistence needs of affected populations in situations of acute risk to survival such as in disaster events. Depending on international relief efforts is justified as it enables people to maintain their livelihood, preventing them from sliding into destitution. Singer's approach cautions about taking the risks of dependence as a justification for withdrawal of relief assistance where acute risks to survival exist. Thus, Singer's approach assumes that only affluent countries and their citizens hold moral responsibility to alleviate suffering and death. However, Sen defends disaster-affected governmental and civil responsibilities to assume such responsibility, calling for relief assistance to incorporate autonomy and freedom for distributed resources to be negotiated and enable affected people to exercise their capacities for deliberation. Such recasting requires that the BNA should expand its list of "basic needs" to include self-help and autonomous choice, without denying the liberating role that external aid may play.

In times of disaster, international aid agencies often build a vulnerability assessment of the affected population for targeting aid distribution. Pogge assumes that relief distribution may lead to stigmatization of those receiving the assistance as passive and dependent. This does not mean halting disaster vulnerability assessment for targeting aid distribution, but rather that international disaster relief agencies need to gain meaningful knowledge of local communities and to assist affected communities to articulate their needs in a way that enables relief agencies to act on them. This also requires international disaster relief agencies to invest in transparent and reliable information of what recipients are entitled to and how assistance is likely to be provided to avoid developing a mentality of dependency syndrome among recipients.

Tackling dependency in international disaster management should operate under principles of equitable distribution of subsistence resources and services to meet immediate needs arising from disaster, while Sen's principle stresses some degrees of freedom and autonomy to allow local communities and NGOs to exercise their ability to establish an informal community support system in the face of a disaster event.

The structure of interdependency provides the basis for considering our shared responsibility to alleviate suffering for "distant" others. Viewed in this way, human freedom derives from the sense of interdependency. Addressing this structure of distributive arrangements helps to lessen the tension between disaster relief and development objectives.

Cooperative forms of disaster relief aid delivery with greater respect for disaster-vulnerable people's needs assessment ought to encourage people who are dependent on external assistance to actively participate in disaster management rather than conceiving of themselves as passive and vulnerable recipients.

5

Donation Fatigue

> There's always the risk of donor fatigue, but that's because need is so widespread. Need is need.
>
> —Howell, 2013

This chapter addresses the issue of donor fatigue, which is caused by budget exhaustion. The growing number and size of natural disaster events may wipe out the donation funding raised by humanitarian aid and relief organizations. In other instances, humanitarian aid and relief organizations may grow frustrated with constant appeals for donations. Frustration may also be intensified when the international community donates to national governments that mismanage loans and donations, resulting in political and administrative corruption, and underdevelopment coupled with high indebtedness.

One of the significant trends in international aid has been declining aid and charitable giving in emerging fields such as natural disasters (Hulme 2010, OECD 2010). The donation fatigue phenomenon can be traced to a development context. Since the early 1990s, with the Cold War effect, several countries and international donor agencies were already falling short on their aid commitments to support developing countries. African countries, which were heavily dependent on foreign aid, have witnessed reduction of poverty relief aid (Bosworth and Collins 1999). "Donation fatigue" was formally adopted by the Tokyo International Conference on African Development (TICAD) in 1993 as a phrase to describe the low levels of motivation on the part of the international community in responding to humanitarian crisis.[1] The TICAD conferences held in Japan were coordinated by the UN (OSSA (Office of the Special Advisor on Africa) and UNDP), AUC, and the World Bank. TICAD was established to focus global attention to pressing issues such as poverty reduction related with African development and aid at the global level between

African leaders and development partners including heads of international and donor organizations. The TICAD initiative aimed at providing a basis for cooperation and partnership relations between Africa and the international community for effective mobilization of information and resources in targeted areas including Promoting Sustainable Economy, Ensuring Human Security including achieving Millennium Development Goals (MDGs) and consolidation of peace, and addressing environment/climate change resilience to natural disasters and violent conflicts.[2] In 1996, United Nations Security Council resolution 1087[3] was adopted as a response to recommendations by the Secretary-General to reduce the mandate of the United Nations Angola Verification Mission III (UNAVEM III) until February 28, 1997, due to signs of donor fatigue regarding the efforts to promote peace and security in Angola.

Despite rich countries' commitments to increase aid for Africa by 2010 at the G8 meeting in Gleneagles in 2005,[4] new net aid flows from G8 countries have not yet increased. Following the World Bank's assessment, net official development assistance (ODA) expenditures overall declined by US$3 billion in 2006 (World Bank 2007, 55). It is claimed that the wars in Iraq and Afghanistan, and the post-9/11 security "imperatives" impact aid flows. Another explanation suggests that aid flows have not reduced but rather have been reallocated to meet new priorities. For example, the United States, which is known as the world's largest supplier of global development aid, was responsible for 25.4 percent of official development aid in 2004/2005 (OECD 2007, Table 8). According to aid figures for 2004, countries of the Near East including Lebanon, Morocco, and the Middle East Region received some US$10 million of ODA while 600 times this amount was spent on other forms of aid such as economic support funds and foreign military spending, which do not fall under the category of development aid as entrenched in the OECD DAC guidelines (OECD 2007). A similar example of diversion of traditional forms of development assistance has been noted in the UK's aid development funding. By 2005, 16.4 percent of total net UK bilateral development aid was contributed to Iraq (as opposed to 0.39 percent in 2002). In conjunction with this evidence, the UK share of multilateral assistance to Iraq accounted for 13.6 percent of all assistance in 2004; by 2005 this had dropped to 4.5 percent (DFID 2007, 263). It is then suggested that while G8 politicians pledge to increase development aid to the poorest countries, these commitments have not yielded increased net aid flows. The impacts of the financial crisis and economic downturn among OECD countries are likely to influence obligations to increase aid flows. This trend is also evident in disaster response funding. For example, while the amount pledged by the government of Indonesia (excluding local government contributions) for the Indonesia tsunami of December 26, 2004, was US$2,100 million, the actual disbursement was US$1,060 million (Tsunami Evaluation Coalition, July 2006, 29).[5]

Evidence of declining foreign aid in developing countries raises the question of why donors have suffered from donation fatigue. According to Collier, decline in foreign aid lies in changed public perceptions: "The key obstacle to reforming aid is public opinion" (2007, 183). Collier suggests that negative public opinion may be a precursor to aid since foreign aid is dependent on supportive public opinion in the long term (Henson, Lindstrom, and Haddad 2010, Mosley 1985, OECD 2003).

The influence of public opinion on support for foreign aid in human psychology causes individuals with low morale and lack of motivation to feel the *need* to act (Barrett and Salovey 2002, Friedrich et al. 1999, Slovic et al. 2002). Psychological experiments demonstrate donation fatigue as a result of emotional drain to motivate intervention to prevent human suffering. The human psychology literature provides evidence on the relationship between perceptions and diminished sensitivity to human suffering, often defined as "psychological numbing." The term *psychological numbing* was coined by Robert J. Lifton (1967) to denote the "turning off" of emotions that enabled rescue workers to effectively function during the relief efforts in the aftermath of the Hiroshima bombing. Psychological numbing refers to cognitive and perceptual factors in evaluating people's unwillingness to fund various interventions (Friedrich et al. 1999).

Slovic et al.'s (2013) model of actual valuation of human lives shows a tendency of lower subjective value of saving lives as the total number of lives saved increases (Slovic, Zionts, Woods, Goodman, and Jinks 2013, 129). According to Slovic et al. (2013, 129), "We may likely not 'feel' much difference, nor value the difference, between saving 87 lives and saving 88." This research indicates a constant increase in the physical magnitude of a stimulus that gradually induces smaller and smaller changes in response.

However, the perception that donation fatigue has increased in recent years is not confined to the development field. The growing number and magnitude of natural disasters throughout the world demonstrate the likelihood that individuals are less willing to donate to humanitarian aid efforts despite the higher number of reported casualties (Slovic et al. 2013, Small, Loewenstein, and Slovic 2007). Zambia, for instance, was faced with massive flooding in 2001, 2005, 2006, and 2007. In 2005, Zambia was also faced with drought. Figures of disaster risk reduction funding show a lower rate of DRR financing despite repeated natural disasters (Kellett and Sparks 2012, 27).

Bangladesh and Pakistan also faced massive flooding between 2000 and 2011, which affected nearly 100,000 million people over this period.[6] However, DRR funding from international donors was not directly related to flooding (Kellett and Sparks 2012, 29).

Many international donors and humanitarian aid agencies perceive that their philanthropic activities for disaster response efforts may not be able to meet needs of the affected population when extreme disasters occur in high proximity and frequency. For example, the 2010 earthquake that struck Chile experienced donor fatigue, which was to some extent caused by the Haiti earthquake, which occurred several weeks prior (January 12, 2010). Although Chile was better prepared for a disaster event than Haiti, the Chilean government struggled to respond to the massive scale of the destruction and appealed for international funding. Donor funding for disaster response efforts was significantly less than in Haiti. While Haiti received $9.3 billion in aid from 2010 to 2012,[7] Chile received $41 million. For their part, international aid actors admitted that their regular donation level fell after the Haiti earthquake.[8]

Drawing on psychological theories and data on aid fatigue confirms the need for the international community to account for the cognitive and psychological constraints described above. As previously noted, cognitive limitations may undermine the mobilization of global community sentiment in the way global justice philosophers advocated as a tool to overcome more obvious constraints such as material and logistical. The field of political economy can sketch the causal dynamics of the global economy and indicate the extent to which donation fatigue could be controlled; theories of global justice provide a metric for evaluating objectives and derive a set of principles with which to approach the problem of donation fatigue in a disaster management setting. In the following section, we differentiate various global distributive justice aspects to consider the implications of donation fatigue on international disaster management.

Analyzing Donation Fatigue in International Disaster Management

What Is Distributed?

Singer's proposal of global charity aims to develop global conscience to directly support a range of worldwide charity projects. Singer suggests:

"The formula is simple: whatever money you're spending on luxuries, not necessities, should be given away" (Singer 1999, 123). Such a formula does push individuals to a broader task of engagement while "[t]hose who do not meet this standard should be failing to meet their share of a global responsibility, and therefore as doing something that is seriously morally wrong. This is the minimum, not the optimal, donation. Those who think carefully about their ethical obligations will realize that—since not everyone will give even 1 percent—they should do far more. But if, for the purpose of changing our society's standards in a manner that has a realistic chance of success, we focus on the idea of bare minimum that we can expect everyone to do, there is something to be said for seeing a 1 percent donation of annual income to overcome world poverty as the minimum that one must do to lead a morally decent life. To give that amount requires no moral heroics. To fail to give it shows indifference to the indefinite continuation of dire poverty and avoidable, poverty-related deaths" (Singer 2004, 194).

Viewed in this way, Singer's ambitious proposal may overcome the cognitive constraints that affect donation fatigue; that is, the deficiencies in people's ability to feel the need to donate for distant others. The act of charity is perceived by Singer as a means to ensure that the cognitive deficiencies would not prevent people from rational calculation and evaluation that cause them to artificially devalue human life: "To give that amount requires no moral heroics. To fail to give it shows indifference to the indefinite continuation of dire poverty and avoidable, poverty-related deaths" (Singer 2004, 194).

In contrast, Pogge rejects ad hoc donations by charitable and affluent public. He insists that long-term engagement with the issue of global poverty demands political mobilization on behalf of global institutional change. Although charity may have the capacity to save the lives of many vulnerable people and communities, it cannot, in itself, be a long-term and sustainable solution to a global regime for economic distributive justice. Charity or donation giving has no mechanism for enforcement or sustainability, while the power and authority of institutions at the global level may have a significant impact on the way resources are distributed to aid the needy. Thus, changing institutional arrangements in the area of international disaster management lies at the heart of aid flows.

Pogge's view may be supported with some evidence from the rise of emerging donors. Due to declining aid donations from the OECD DAC (Development Assistance Community) members, countries such as China, the United Arab Emirates, Saudi Arabia, South Korea, Venezuela, India, Kuwait, and Brazil have been increasing their aid donations to poor countries. These countries, also termed "emerging donors," are enormously generous and provide aid donations with no demands that the recipient countries should work to improve good governance.[9] Donation aid with no conditionality may jeopardize existing international institutional arrangements and ethical standards that aim to fight corruption in aid delivery and improve transparency and accountability.[10] More subtly, a seeming absence of conditionality in development aid may cause poor countries to turn down aid that comes from international agencies bound to multilateral standards. China, for example, has moved beyond the World Bank anti-corruption mechanism to fight corruption in Nigeria's transportation sector by offering an unconditional loan to fund railways in Nigeria (Naim 2007). Similarly, in Indonesia, China offered to expand the country's electrical net by building plants that do not meet anti-pollution standards. In 2005 Angola halted the IMF efforts to introduce a staff-monitored program to oversee Angola's economic policies, an act that was supported by a $2 billion package of soft loans from China (Pawson 2007).

Sen also discerns the politics of aid flows by claiming that differences in donor behavior can be ascribed primarily to differences in the underlying social and political circumstances. In addition, evidence shows that aid tends not to reach the most vulnerable sectors in society, thus providing a compelling reason to distribute capabilities to expand people's opportunities to pursue their goals. Donation fatigue, which is often born out of frustration that aid has not met its promises, justifies the need to amend aid distribution in a way that removes obstacles in people's lives to enable them to gain more freedom to live the kind of life which they have reason to value (Sen 1992). "In getting an idea of the well-being of person, we clearly have to move on [from commodities and characteristics of commodities] to 'functionings,' to wit, what the person succeeds in doing with the commodities and characteristics at his or her command. For example, we must take note that a disabled person may not be able to do many things an able-bodied individual can, with the same bundle of commodities" (Sen 1999a, 10).

Community of Justice

Research on donor behavior and psychological numbing in the context of disaster events often points to the fact that people are more willing to donate to or aid identified vulnerable individuals rather than unidentified or statistical victims (Jenni and Loewenstein 1997, Kogut and Ritov 2005a, Schelling 1968, Small and Loewenstein 2003, 2005). However, in a psychological experiment conducted by Small, Loewenstein, and Slovic (2007), people were given the opportunity to contribute up to $5 of their earnings to the Save the Children organization. In one condition respondents were asked to donate money to feed an identified victim, namely a seven-year-old African girl named Rokia. The figures showed that this group contributed more than twice the amount given by a second group asked to donate to the Save the Children organization, which helps to save millions of Africans from starvation (Small et al. 2007).

The respondents in a third group were asked to donate to Rokia, but were also shown the number of victims shown to the second group. Coupling the dry statistics with Rokia's story significantly *reduced* the contributions to Rokia. The researchers admitted that the presence of statistics might have reduced the attention given to the story of Rokia that was essential for establishing the emotional sympathy necessary to motivate donations. Other studies on the effect of identifiable and statistical victims on donors' motivation to donate yielded controversial inclinations. For example, Kogut and Ritov (2005a,b) studied differences in contribution between identified victim and identified group. They showed that people tend to feel more compassion when regarding an identified single victim than when regarding a group of victims, even if this group is identified. The results showed that contributions to the individuals in the group, as individuals, were far greater than were contributions to the entire group. Reported feelings of distress were also higher in the individual condition. Kogut and Ritov then concluded that the greater donations to the single victim most likely result from the stronger emotions evoked by such victims (Kogut and Ritov 2005b).

In general, these studies suggest that motivation to donate and aid victims in large-scale disaster events may be hard to arouse and sustain over time for a larger number of victims and be even less for unidentified groups. Thus, cosmopolitan justice theories may find it hard to justify a sizable capacity and attention to care for other distant and unidentified victims. For instance, Singer advocated an extension of the community of justice beyond the boundaries of a given neighborhood, ethnic community, or nationality as grounds for choosing to give charity to one person rather than another: "It makes no moral difference whether the person I help is a neighbor's child ten yards from me or a Bengali whose name I shall never know, ten thousand miles away" (Singer 1972a, 231–32).

Although Singer's position on the community of justice seems emotive to bring more people within the ambit of moral concern, and appealing when it comes to immediate relief efforts needed for a large scale disaster, it may still fail to motivate action for distant and unidentified victims. As the intensity of a large-scale disaster increases, donors may feel that the dynamics of such events are too extreme and that their individual efforts are meaningless, which eventually leads them to decrease or even cease their giving.

Contrary to Singer, Pogge does not underestimate the power and authority possessed by states and international agencies to constrain and enforce their members to provide aid since these organizations are held accountable for their actions. Due to historical circumstances, states assume more control over the international economy, allowing them to choose where their imports and exports are allocated: "The presence and relevance of shared institutions is shown by how dramatically we affect the circumstances of the global poor through investments, loans, trade, bribes, military aid, sex tourism, culture exports, and much else" (Pogge 2001, 61). Control over the international market can also yield institutional solutions to ensure access to basic resources at the global level. Thus, when it comes to distributive issues, it is only laws and institutions that take members' interests into account—be it members of religious, ethnic, or national groups. Global community therefore includes nation-states and other agencies of corporate response such as corporations, NGOs, and IGOs.

Sen's perception of members of the community of justice addresses differences in donors' and recipients' behavior. Behind poor countries and individual aid recipients are people who are neither powerless nor ignorant with respect to problems and opportunities for relief efforts; they need to be considered as agents of change, capable of independent as well as cooperative action: "I am using the term agent . . . in its older— and 'grander'—sense as someone who acts and brings about change, and whose achievements can be judged in terms of her own values and objectives" (Sen 1999a, 19). "The people have to be seen . . . as being actively involved—given the opportunity—in shaping their own destiny, and not just as passive recipients of the fruits of cunning development programs" (Sen 1999b, 53). However, it should be noted that this perception of heterogeneity of human beings' capabilities and their goals may result in weak motivational forces for aid on the part of the donors who may feel that differences between recipients can be worthless, which eventually decreases aid flows.

Preconditions of Just Distribution

The preconditions of just distribution at the global level rest, according to Singer, on the plight of those whose substantive needs remain unfulfilled, to the immediate danger of their lives and communities. Thus, Singer rejects Rawls's assumption of the legitimate conditions of global coercion (among states): "As a result, the economic concerns of individuals play no role in Rawls's (sic) laws for regulating international relations. . . . As our world is now . . . millions will die from malnutrition and poverty-related illnesses, before their countries gain liberal or decent institutions and become 'well ordered'" (Singer 2004, 179).[11] Under the conditions of complex interdependence of the global world, Singer holds that there is a greater probability for direct well-targeted donations by individuals and governments rather than for major structural reform (Singer 1972b, 114–16).

By contrast, Pogge draws on the historic conditions that led to the severe global economic inequality reflected in the existing structures of the global economy. Thus the key to maintaining sustainable foreign aid flows, changing the political economy of OECD countries, along with changing economic priorities will make them more willing to give aid. For example, during the mid-1990s, European countries experienced high levels of unemployment along with demographic changes that have led to increasing economic pressure on domestic elements of government expenditure. Such evidence made the political climate much less supportive of giving aid or financial donations, which become costly in terms of domestic perspective (Spektorosky and Ireni Saban 2013).

Sen also recognizes of the role of political and economic conditions behind global distribution in supplying freedoms with a conscious commitment. Such commitment compels donors to use greater selectivity in the allocation of aid, namely, to adjust their donation to include an appropriate set of incentives for long-term resilience plans chosen by disaster-affected populations. An appropriate incentive structure that follows "the freedom to achieve actual livings that one can have reason to value" (Sen 1999b, 73) will be less susceptible to donation fatigue.

THE PRINCIPLE OF DISTRIBUTION (CONTENT)

Singer's mutual assistance principle, according to which people in affluent countries are morally obligated to help those who are unable to meet their subsistence need (Singer 1972a, 231), aims at ensuring the mutual advantage of recipient and donors. Global distribution according to need may be adopted as a legitimate distribution principle if survival is at stake. This principle takes into account the extent to which people's life chances with a state are affected by structures of social inequality and events beyond the national boundaries and control. Thus, Singer's principle helps to distinguish between targeted donation and institutional reforms. However, in adverse situations such as extreme disaster, we may make immediate choices that can result from making unwittingly suboptimal compromises. Singer himself admits: "Caring about doing what is right is, of course, essential, but it is not enough, as the numerous historical examples of well-meaning but misguided men indicate" (Singer 1972b, 4).

Pogge extends the Rawlsian difference principle at a global level by recognizing that some inequalities can be justified—on the basis that they improve the conditions of the most needy or of all members of the global community. Pogge's principle provides the criteria for distinguishing fair from unfair asymmetric distributions, which is missing from Singer's theory. Sen's principle is also structure-sensitive, and leads to considering the political and economic consequences for assessing peoples' capabilities. According to Sen, the principle of justice requires basic capability equality so that each person equally has freedom to attain a sufficient level of basic capabilities. By focusing on capabilities, and on the ways in which others are vulnerable to us, we should take into account the context and life circumstances of an individual agency. Vulnerable people are neither powerless nor ignorant with respect to important problems and opportunities for action; they need to be addressed as agents, capable of independent action as well as cooperative endeavor. Agency is conceived in terms of responsible autonomy, an others-regarding way of deciding and acting (Sen 1999a, 9). Thus, donor recipients who consider their capabilities relative to other members of the global community are more effective in coordinating collective action and better at mobilizing their aid needs.

THE PRINCIPLE OF DISTRIBUTION (STRUCTURE)

Singer's mutual assistance principle, which obligates affluent individuals to donate their excess wealth, is based on volunteerism with no mechanism of enforcement behind it. The structure of volunteerism based on impartial and universal structure follows the idea that aid has a humanitarian role, meaning that even if it is ineffective in reducing inequality between countries, it could be successful in raising the awareness of the plight of individuals who are dying of starvation and malnutrition across the globe.

However, Pogge is concerned that if just obligation will not be supported by institutions and global economic structures, aid will fail to be of sufficient size or duration to adequately address problems of global poverty and also be more susceptible to donation fatigue. Sen's proposal for a local parochial structure behind the principles of justice to maintain open impartiality can be seen as a departure from both Singer's and Pogge's structures. According to Sen: "This open conditionality [of the responsible agent] does not imply that the person's view of his agency has no need for discipline, and that anything that appeals to him must, for that reason, come into the accounting of his agency freedom. The need for careful assessment of aims, objectives, allegiances, etc., and of the conception of the good, may be important and exacting" (Sen 1985, 204). The structure of open impartiality suggested by Sen conceives global poverty reduction as multidimensional. That is, it acknowledges the fact that there can be more than one objective (knowledge, health, work participation) that holds an intrinsic value in a given society, and that the set of valued ends and their relative weights will vary across individuals and cultures. But if individuals' ends are so heterogeneous and cannot be adequately responded to by a common measure such as charity or income, this creates a problem of targeting aid donation for both donors and recipients.

Summing Up

In this chapter we discussed the ethical and moral considerations, along with the cognitive and perceptual constraints, imposed on meaningful provision of aid or donation. The phenomenon of donation fatigue amongst the global community members requires more than moral intuitions to motivate donors to make a proper response effort. This places the burden of response on global distributive justice approaches. Singer defends the value of charity benefitting from the meaningful engagement of volunteers. Since bringing about institutional reforms takes time and long-term commitments, volunteering on an ad hoc basis can exert influence over public opinion and moral behavior.

Pogge's and Sen's approaches are more structure sensitive in terms of indirect and long-term impact of aid and relief actions on recipients and the global economic system as a whole. A structural account constantly requires us to create laws and institutional arrangements to help overcome the psychological deficiencies in our ability to feel compassion and to act for the needy. Although Singer's proposal of charity may well have the capacity to save many lives—while it may be precisely what we need in adverse situations—until we are able to develop a global regime for sustainable aid flows—it is not in itself an adequate grounding for global distributive justice. The rise of "emerging donors" who provide aid donations with no conditionality addresses the significant role of institutional arrangements and standards in aid donation. A global regime for sustainable aid flows must ensure that emerging donor governments, private companies, and NGOs are all committed to standards that multilateral actors establish. From Pogge's institutional approach to global justice, the concern of the increase of emerging donors reveals important deficiencies in the existing system of development finance that should be tackled.

The problem of donation fatigue often highlights the way donors' interests and priorities have dominance over recipients. Thus, Sen warns against the image of both Singer and Pogge of recipients as individuals or communities in developing countries. Poor individuals and communities are neither powerless nor passive but rather active recipients capable of independent actions and cooperative efforts commensurate with the value they place on their lives and goals. Viewed in this way, declining aid can be seen as an entirely positive thing that motivates affected individuals to employ their capacities and power independently.

6

Corruption

> Humanitarian assistance aims to save lives and alleviate the suffering of people in times of crisis. Yet these noble ambitions do not immunise emergency responses from corrupt abuse.
>
> —Transparency International 2010, viii

In disaster events, humanitarian aid involves large amounts of cash and supplies that may create opportunities for corruption, especially in developing or corrupt countries. This chapter discusses the way corruption in the allocation of aid undermines the moral obligation behind humanitarian assistance, leading to inequitable and ineffective distribution of disaster relief and reconstruction aid. More specifically relevant to the context of global distributive justice, corruption occurring within disaster-affected countries may push donors to cut back on their aid funding. The threat of waste and corruption by disaster-affected governments poses distributive consequences, particularly reputational consequences for domestic agencies, and a decline in aid funding for lack of the authority to convincingly deliver international aid to those in need. Corruption risks may undermine international funding for disaster mitigation efforts in the donor country, with taxpayers being legitimately concerned that their funds are being stolen rather than assisting disaster-affected citizens.

Corruption is defined as the "misuse of public power for private gain" (Dobel 1978, Nye 1967, 419). Such a definition of corruption includes legal, illegal, and borderline actions, which can allow the legal system to enforce sanctions for unethical action. However, if a society defines certain behaviors as not corrupt, but the legal system defines them as corruption, the legal system will be unable to enforce legal standards of ethical behavior. Thus, an alternative definition of corruption based on public opinion perspective has been suggested by Heidenheimer (Heidenheimer 2002, Heidenheimer et al. 1989). The Heidenheimer definition of corruption has the advantage of being conceived in social terms as what the public—in any given society—sees as being corrupt. The indicator for corruption is deviation from the law, which serves as the value guiding the legislators and judicial elite. This perception plays a part in public opinion regarding political corruption. In order to bring the subject of corruption to public awareness, Transparency International (TI), the largest worldwide organization comparing levels of corruption, publishes annually the "Corruption Perception Index." This index determines the level of corruption in each country based on the perceptions of businessmen, academics, and foreign and local risk-analysts. According to this definition, the level of corruption of a certain act is determined by the result of the combined judgment of the elite and the general public.

Alternative definitions of corruption has been suggested by Kurer (2005) and Rothstein and Teorell (2008). Their definition is broadened to include violations of professional norms and ethics, compared to the narrower legal and public opinion definition of corruption. According to Kurer (2005, 230) and Rothstein and Teorell (2008), corruption "involves a holder of public office violating the impartiality principle in order to achieve a private gain." Such a definition is based on an ethical-normative approach that addresses the role of professional ethics and guides elected officials' practice. In general, the purpose of each profession is to promote a particular, valued goal of great importance to people's well-being and to follow principles and rules to guide proper behavior (Kasher 2003). While doctors properly aim at the health of their patients; lawyers, at legal justice for their clients; and teachers, at the education of their students; we say that each of them performs professional acts within professional practices. The understanding of the essence of professional practice is the major aspect in distinguishing one profession as different from all others. For example, the medical and nursing professions are different, yet they share the same value of healing. They must be differentiated on the grounds of specific rules and principles in the pursuit of the patient's health.

These principles and rules should spell out the nature of the professional practice of a given profession. For example, political corruption is viewed as a violation of specific normative-ethical principles underlying elected officials' practice, namely the impartiality principal (Frederickson 1993). To act in an impartial way is "to be unmoved by certain sorts of consideration—such as special relationships and personal preferences. It is to treat people alike irrespective of personal relationships and personal likes and dislikes" (Cupit 2000). Impartiality concerns the nature of moral deliberation aimed at removing the biasing influences of one's objectives, interests, and favoritism based on the agent's personal characteristics, background, values, and beliefs (Brighouse and Swift 2009, Cottingham 1986, Scheffler 2001, Williams 1974).

Corruption is often blamed for the failures of certain "developing" countries to achieve economic growth. According to Axel Dreher and his colleagues (2007), corruption is conceived as one of the causes for a reduction of 58 percent in per capita income in Latin America and the Caribbean, and 63 percent in sub-Saharan Africa and south Asia. Fedderke and Klitgaard (2006) suggested that, other things being equal, countries with more corruption have less investment, and each dollar of investment has minor impact on economic growth (Fedderke and Klitgaard 2006, Klitgaard 2006, Klitgaard et al. 2005). Kaufmann et al. (2009) have shown that measures of poor governance, such as high corruption, lack of rule of law, and lack of transparency, have direct and negative effects on long-term outcomes such as infant mortality and educational attainment. At the same time, corruption is seen as one of the main barriers faced by post-communist countries in reaching the stage of democratic institutions (Rothstein and Teorell 2008).

Given the current debates regarding inflows of aid resources during large-scale disasters, these approaches that address the effects of corruption on development and democratization processes are particularly timely as institutional development is thought to be a dependent variable, affected by targeted aid (Callaghy 1988, Chabal and Daloz 1999, Sandbrook 1992, van de Walle 2001). Therefore, corruption is often defined as a development challenge (Holmberg, Rothstein, and Nasiritousi 2009). An "aid-corruption paradox" emerges when the need for foreign aid is higher in corrupt countries while corruption undermines the moral value behind foreign aid. For example, according to the 2010 Corruption Perceptions Index, four countries that require extensive foreign aid for their development, namely, Somalia, Myanmar, Afghanistan, and Iraq (Transparency International 2010), are highly corrupt countries. These figures stress the fact that need for foreign aid is often greater in more corrupt countries. This understanding of the aid-corruption paradox is common in the international disaster management setting. During disaster events, huge inflows of foreign aid may pave the way for corruption. In these settings, corruption can intensify disaster vulnerability by government misuse of loans and cash donations from the international community, failing to invest in emergency management activities or to spend on disaster mitigation or preparedness projects, and may also lead to inequality in humanitarian assistance and relief distribution. In addition, corruption in the public sector may intensify the damage caused by natural disasters because corruption can lead to reduced building safety, and therefore reduced disaster resilience.

This was the case during Japan's 2011 devastating earthquake, which led to leakage of the Fukushima Daiichi nuclear plants. The Fukushima accident resulted from the combination of natural and human risks. Japan has a contracted electricity market, with a large share of the market held by a few providers. For that, electricity companies in Japan hold an extended market power and gain greater profits than in a competitive market. Corrupt government regulators enable the companies to overcome administrative barriers such as safety standards for the purpose of providing monopoly profits for favored producers (Bliss and Di Tella 1997, Escaleras et al. 2007). The Tokyo Electric Power Co., which operates the Fukushima Daiichi nuclear plants, ignored warnings that the reactors were at risk when faced with natural disaster. However, the Tokyo Electric Power Co. was privileged by government officials. "Just a month before a powerful earthquake and tsunami crippled the plant at the center of Japan's nuclear crisis, government regulators in Japan approved a 10-year extension for the oldest of the six reactors at the power station 12 despite warnings about its safety." (Tabuchi et al. 2011). The risk of a catastrophe had been appropriately predicted, which finally resulted in massive amounts of radioactive water, continued spills of contaminated water at the plant, and massive evacuation (Hammer 2011).

The Gulf Coast relief and response efforts in 2005 exemplify the way government agencies' poor management in meeting immediate needs after a disaster strikes can lead to corruption. Consider, for example, the case of contract abuse in post-Katrina response efforts. In the aftermath of Hurricane Katrina, the Bush Administration turned to private contractors to offer relief and recovery services up to billions of dollars.[1] These Katrina-related contracts were set with inadequate contract management and government oversight, leading to waste, fraud, and abuse in federal contracting.[2] Indeed, government reports and audits have documented deficiencies in contract management at both the Federal Emergency Management Agency (FEMA) level and the Army Corps of Engineers, the two agencies that were entrusted with the responsibility for most Hurricane Katrina recovery effort. The justification for the award of noncompetitive contracts during the Hurricane Katrina response efforts as the percentage of contract dollars awarded without full and open competition actually increased months after Hurricane Katrina. For instance, in September

2005 (a month after Hurricane Katrina) 51 percent of the contract dollars awarded by the Federal Emergency Management Agency were awarded without competitive bidding. By June 30, 2006, over $10.6 billion had been awarded to private contractors for Gulf Coast recovery and reconstruction. Over 47 percent of these contracts were awarded without full and open competition.[3]

Continued evidence of fraud, waste, and abuse were found within FEMA's Individuals and Households Program (IHP) and in the Department of Homeland Security's (DHS) purchase card program.[4] In the aftermath of both hurricanes (Katrina and Rita) FEMA made nearly $20 million in duplicate payments to thousands of individuals who claimed damages to the same property from both hurricanes Katrina and Rita. FEMA had also disbursed millions in potentially improper and/or deceptive payments to nonqualified aliens who were not eligible for IHP. By 2006, the GAO estimated that an estimated US$1.5 billion of FEMA's expedited assistance (EA) payments were deceptive.[5]

Thus, governance failures that abet corruption—such as inefficient administrative structure, inadequate planning and preparation to anticipate requirements for disaster needed goods and services, a lack of clearly communicated responsibilities across agencies and jurisdictions, inadequate deployment of personnel to support an effective disaster management system, and limited controls—become more prominent in disaster settings. The link between governance and corruption is also evident in the case of aid delivery to Haiti following the 2010 earthquake. In the aftermath of the devastating earthquake in 2010, Haiti received unprecedented aid for relief and recovery efforts.[6] However, the hefty aid did not yield any achievements in terms of its economic and disaster management domains with aid donations gone awry given the country's legacy of dysfunctional bureaucracy: "With the earthquake, you essentially had billions of dollars being sent into that very broken, top-heavy structure, which was made far worse by this massive infusion of many more NGOs" (Schuller 2012). Bureaucratic aberrations and doubtful behavior on the part of politicians fostered distrust and cynicism among Haitian citizens: "Despite the best intentions of the international community Haitians have little faith they will see the billions of dollars in aid pledged to rebuild their earthquake-shattered country, which international monitors rate as one of the world's most corrupt" (Zengerle 2010). Increased concerns over the rise in Haiti's drug economy resulted from the chaos following the earthquake: "Haiti has always been a weak link against drug trafficking.

It's a grave situation, and it's going to get graver, because people are now going to be even more susceptible to whatever corrupting forces are out there" (Hawley 2010).

Analyzing Corruption in International Disaster Management

What Is Distributed?

In the postscript to "Famine, Affluence, and Morality," Singer draws a distinction between development assistance and humanitarian aid. According to Singer, by distributing development assistance, we fulfill our obligation to the poor as it "is usually the better long-term investment" (Singer 1972, 241). On the other hand, humanitarian aid is associated with resources (e.g., food, shelter, and medical care) necessary to alleviate immediate suffering or death. Corruption negatively impacts development assistance humanitarian aid; there is little reason to be optimistic about making development assistance more effective in reducing poverty. Thus, development assistance should be motivated by purely strategic thinking of alleviating global poverty while immediate aid to disaster-affected corrupt countries should follow Singer's urgency of our duty to provide immediate aid to save lives even if such aid is at best a minor part of what we should do to reduce suffering. Pogge also believes that development aid is often ineffective in reducing poverty when viewed from a political economy perspective: "In some poor countries, the rulers care more about keeping their subjects destitute, uneducated, docile, dependent and hence exploitable. In such cases, it may still be possible to find other ways of improving the circumstances and opportunities of the domestic poor: by making cash payments directly to them or to their organizations or by funding development programs administered through United Nations agencies or effective non-governmental organizations. When, in extreme cases, GRD funds cannot be used effectively in a particular country, then there is no reason to spend them there rather than in those many other places where these funds can make a real difference in reducing poverty and disadvantage" (Pogge 2001, 68). Once disaster occurs, existing patterns of corruption do not fade away. Thus, the right to security as part of the right of subsistence ensures that a disaster-affected population can meet its basic needs, as it is our duty to compensate for the inequality and corruption that caused victims in corrupt countries to have less ability to fight corruption. Since Sen conceives development as a multidimensional process of expanding a set of freedoms (Sen 1999a, 3, 37, 53), it cannot be jump-started by development assistance. This entails a more intrinsic evaluation of freedoms to achieve the development objective rather than instrumental effectiveness: "The intrinsic importance of human freedom as the preeminent objective of development has to be distinguished from the instrumental effectiveness of freedom of different kinds to promote human freedom" (1999a, 37). Corruption can be seen as one dimension of poverty in developing countries, and is neither the sole content nor leading cause of it. Seen in this light, humanitarian aid in disaster events should go beyond the dimension of subsistence, such as resources that enhance affected-community resilience. Enhanced resilience allows better anticipation of disasters and better planning to reduce disaster losses, namely development of partnerships and networks to build resilience at the government and local community levels, preparation and provision of guidelines, information, and other resources to support community resilient efforts.

Community of Justice

Singer's criteria of membership in the global community include the capacity of suffering or having preferences. In the context of corruption, all members may well realize the higher costs of being caught rather than the expected gains from corruption and prefer to refrain from corrupt practices: "From a utilitarian perspective, punishing those guilty of past crimes will, one hopes, put others who might do something similar on notice that they will have no refuge from justice, and so deter them from committing new crimes" (Singer 2004, 120). However, both Singer and Pogge agree that the only reason they would refrain from corrupt behavior is if institutions could be established to maintain trust that most other agents would refrain from corrupt behavior as well (Singer 2004, 127). Thus, in our case, behavior depends on how the existing institutions inform the agents' mutual expectations; that is, the expectations of whether the other agents will take part in corrupt behavior. Pogge specifies that the capacities needed to engage in the community of justice from an institutional perspective must ensure that "they are freed from bondage and other relations of personal dependence, that they are able to read and write and to learn a profession, that they can participate as equals in politics and in the labor market, and that their status is protected by appropriate legal rights that they can understand and effectively enforce through an open and fair legal system" (Pogge 2001, 68). Sen seems to doubt whether the establishment of impartial institutions that maintain the agents that most of "the others" will refrain from corrupted behavior, especially in a society marked by systemic corruption in which members of the community of justice are deprived of choosing whether to take part in a corrupt exchange. Such capacity for "reasoned evaluation" (Sen 1999a, 78) is done by members of the community of justice through rationality. "Rationality is interpreted here, broadly, as a discipline of subjecting one's choices—of actions as well as of objectives, values and priorities—to reasoned scrutiny" (Sen 2002, 4).

Hence, the task of international aid organizations to reduce disaster vulnerabilities in corrupt countries cannot be fulfilled without an understanding of these functions. The role of international disaster management is to provide external opportunities by encouraging vulnerable populations through education and through the local community to look for internal capabilities and avoid institutions that could block these capabilities.

Preconditions of Just Distribution

Since aid is seen by Singer as an altruist undertaking, donors should be identified with having a positive duty to assist those less fortunate even in corrupt recipient countries. Thus, concerns of corruption are regarded as "more practical than philosophical" (Singer 1972, 231) as they are not of comparable moral import to the actual lives that can be saved simply by donating to humanitarian aid organizations. The conditions of just distribution of aid at the global level are not part of the underlying economic systems in corrupt countries but rather account for the level of suffering and vulnerability in these countries faced with adversity. Moreover, focusing on corruption effects in the recipient countries can give rise to donation fatigue and thus increase suffering and death. "Granted in normal circumstances, it may be better for everyone if we recognize that each of us will be primarily responsible for running our own lives and only secondarily responsible for others. This, however, is not a moral ultimate, but a secondary principle that derives from consideration of how a society may best order its affairs, given the limits of altruism in human beings. Such secondary principles are, I think, swept aside by extreme evil of people starving to death" (Singer 1972, 239).

Sen aims at expanding Singer's conceptualization of moral preferences as conditions of just distribution by referring to an adaptation process over time, from ongoing interactions between members of society.

Adaptation is defined by Sen as a mental state of accepting the traditional and habitual circumstances as an effect of an adverse social and economic environment that influences the agent's perceptions: "The destitute thrown into beggary, the vulnerable landless labourer precariously surviving at the edge of subsistence, the overworked domestic servant working round the clock, the subdued and subjugated housewife reconciled to her role and her fate, all tend to come to terms with their respective predicaments" (Sen 1985, 21).

Adapted preferences concerns the way a country confronts and manages corruption or other adversities that in turn influence adaptation to all members and their relationships. Sen, similar to Singer, does not focus narrowly on the destructive cycle of corruption and its impact on distributional arrangements, but rather attends to its context. Thus, in assessing the conditions for aid distribution in corrupt countries, it is essential to explore how recipient country members handle corruption incidents, their proactive stance, and long-term "survival" strategies.

Pogge also holds that political and economic conditions such as corrupt elites and bureaucratic corruption in developed countries do not absolve us of responsibility to provide them aid. Pogge's account of global inequality dismisses any excuses of the developed world for not giving aid or supporting reforms of the global order. Indeed, the developed countries have created the global order and benefit from it disproportionately to the developing world (Pogge 2002, 13). Under these conditions the developed countries are met with negative obligations to reform the institutions so that they erase harming effects to other countries as "the distinction between causing poverty and merely failing to reduce it [is] morally significant" (ibid.).

THE PRINCIPLE OF DISTRIBUTION (CONTENT)

Singer's mutual assistance principle, which applies the positive duty of affluent countries to assist those who are unable to meet their needs in the face of adversity, tackles the relationship between corruption and aid fatigue that may lead to lessening support for foreign aid: "For each of us there will be many things on which we spend money that we do not truly believe to be of comparable moral importance to death by starvation" (Singer 1999, 303). Singer explains the utilitarian rationale behind the principle of justice: ". . . even if a substantial proportion of our donations were wasted, the cost to us of making the donation is so small, compared to the benefits that it provides when it, or some of it, does get through to those who need our help, that we would still be saving lives at a small cost to ourselves—even if aid organizations were much less efficient than they actually are" (Singer 1997, 28).

Pogge offers a solution for developed nations to pursue that will "protect [institutional order] victims and promote feasible reforms that would enhance the fulfillment of human rights" in his proposal of a "Global Resources Dividend" (GRD) (Pogge 2002, 196). A global resource dividend would compensate for resource privileges and unfair trade practices. Pogge views the GRDs as a tool that would reduce the incentives for corrupt leadership in developing countries. The GRDs would mitigate the adverse consequences of the existing institutional order to developing nations.

Pogge proposes the GRD as a prerequisite for states and their governments to "share a small part of the value of any resources they decide to use or sell," and the proceeds of a GRD will be used to increase the access people have to the basic goods needed for minimal human flourishing. The goal, then, is for all persons to "effectively defend and realize their basic interests" (ibid.).

An example of how the GRD framework should be effective in preventing corruption during disaster response and relief actions is the Anti-Corruption Agency (ACA) activities in Malaysia during the 2004 Indian Ocean Tsunami (Mohamad 2005). On December 26, 2004, when the Indian Ocean tsunami hit the states of Penang, Perlis, Kedah, and Perak in Malaysia, existing anti-corruption institutional mechanisms were applied. In 1961, the Malaysian government had initiated an independent Anti-Corruption Agency (ACA) to enforce the Malaysian Prevention of Corruption Act. In 1998, Integrity Management Committees (IMCs) were instituted across federal and state agencies. In the aftermath of the tsunami, when the National Disaster Aid Fund began a process for managing the RM 90 million (US$24 million) for disaster relief, ACA Penang had already deployed its corruption prevention system. Once the police had finished their affected-persons' loss and property damage report, three separate state committees, each with elected and local community representatives, as well as other government entities, reviewed these reports before they were submitted to the National Disaster Aid Fund Management Committee for final approval. Other anti-corruption measures were taken such as the publication of assistance amounts, providing information on the assistance at the time of disbursement, and requiring that the government official and the recipient sign a form that informs of the consequences of providing deceptive information.

By drawing on the principle of equal capabilities, Sen finds that in most cases, members of corrupt recipient counties tend to be ill-informed of the exact amount of foreign aid and how it is distributed. Following Sen's principle of justice, four central constructs are linked: recipient engagement in active process to balance local needs with local capabilities as these interact with local evaluations and meanings to arrive at a level of community resilience adaptation in the face of adversity. Thus, competent functioning (resilience) can lessen the impact of corruption on donation after exposure to disaster in a corrupt country: "The people have to be seen . . . as being actively involved—given the opportunity—in shaping their own destiny, and not just as passive recipients of the fruits of cunning development programs" (Sen 1999a, 53). Policy design should then aim at reducing vulnerabilities of members of corrupt countries so that they have more freedom to handle adversities as well as to deal with development objectives.

THE PRINCIPLE OF DISTRIBUTION (STRUCTURE)

Singer's universal structure of the principle of justice aims at ensuring that people act in an impartial way when considering donation to disaster-affected countries. The fact that corruption becomes a reason for lessening support for foreign aid (Erixon and Sally 2006) supports Singer's claim for impartiality; that is, that motivations for distributing aid will not be mediated by perceptions of corruption. Pogge offers to overcome systemic corruption through impartial institutions. International institutions will ensure through the agent that most of "the others" will fulfill their obligations. Viewed in this way, impartial institutions need to provide information of "mutual expectations." This structure reinforces arguments for international humanitarian organizations to invest more in improving the transparency of aid distribution to corrupt countries. According to Pogge: "Once the agency facilitating the flow of GRD payments reports that a country has not met its obligations under the scheme, all other countries are required to impose duties on imports from, and perhaps also similar levies on exports to, this country to raise funds equivalent to its GRD obligations plus the cost of these enforcement measures. Such decentralized sanctions stand a very good chance of discouraging small-scale defections" (2001, 69). This idea, institutional initiatives concerning curbing corruption in the face of natural disaster, can be found in the G8 Summit, which was held in France two months after Japan's 2011 disaster. Since Japan's disaster had shaken citizens worldwide, the members of the G8 Summit supported the need for more stringent international rules on nuclear safety following the disaster at Japan's Fukushima plant. French President Nicolas Sarkozy said: "We all wish to get a very high standard of regulation on nuclear safety, that will apply to all countries involved in civilian nuclear energy and which will take safety to the highest levels ever."[7]

Sen seems to doubt the effectiveness of impartial policies and institutions in countries that suffer from systemically corrupt structures. Reforms to improve transparency and accountability in aid delivery to corrupt countries should not be confined to formal institutions but also to the informal institutions in recipient countries. The quality of informal institutions in the recipient society reflects the moral standards of that society due to the fact that "there are strong influences of the community, and of the people with whom we identify and associate, in shaping our knowledge and comprehension as well as our ethics and norms. In this sense, social identity cannot but be central to human life" (Sen 1999b, 5). This local evaluation is often neglected from corruption assessment by foreign organizations; therefore Sen's open impartiality structure places conditions on aid delivered to a corrupt society, such as responsibility: "This open conditionality [of the responsible agent] does not imply that the person's view of his agency has no need for discipline, and that anything that appeals to him must, for that reason, come into the accounting of his agency freedom. The need for careful assessment of aims, objectives, allegiances, etc., and of the conception of the good, may be important and exacting" (Sen 1985, 204).

Summing Up

In recent years, the international community has questioned the efficacy of international humanitarian aid delivered to corrupt countries affected by large-scale natural disasters. Corruption in recipient countries may lead to reduced support for international humanitarian aid. Such dilemmas need to be addressed by theories of global justice as aid distribution to corrupt countries is mediated by perceptions of fundamental moral imperatives. All three philosophers hold that aid helps developing countries with much needed resources in the face of adversity. The question remaining is how to ensure that waste in aid may not lead to public concern of aid effectiveness, which often results in donor fatigue.

Singer suggests that the influence of perceptions of corruption in general on support for international aid should not affect motivation to donate by pointing at the possible concerns of aid fatigue in a world faced with continuous problems of poverty and death. Pogge proposes an institutional solution to fight corruption in developing countries by improving the transparency and accountability of aid delivery and distribution established by impartial institutions in the complex web of interdependencies at the global level.

Sen doubts the capacity of impartial institutions alone to fight corruption in disaster aid delivery as supported by both Pogge and Sen. Sen agrees with the practice of imposing conditions on aid as a viable option for effectiveness of aid delivery to corrupt countries; however, he emphasizes the role of informal institutions at the local level, which can lead to raised awareness and empower citizens to develop new and renewed competencies, and build mutual support and collaborative efforts among affected community members. By strengthening community resilience in corrupt settings, international aid organizations can build local resources to meet disaster vulnerabilities more effectively. International programs and projects need to be developed proactively to meet emerging global challenges, including aid fatigue; therefore international organizations must help local communities avert breakdown and seize opportunities for resilience and recovery out of disaster. In addition, inadequate preventive controls related to the disaster aid funding management process, which resulted in an estimated $1 billion of potentially improper and/or fraudulent payments in the aftermath of 2005 Hurricanes Katrina and Rita, supports the need to implement institutional arrangements of fraud prevention, which is far more effective and less costly than detection and monitoring, given the urgency of meeting needs in the face of adversity.

7

Compensation

> ... [C]ompensate for the damages we cannot prevent, and prevent the damages for which we cannot afford to compensate.
>
> —Aubrey Meyer, of the Global Commons Institute. Meyer's original statement was "adapt to the changes we cannot prevent, and prevent the changes to which we cannot adapt."
> In conversation, May 12, 2007

The devastating typhoon that caused the death of thousands of people in the Philippines has reinvigorated the debate about whether rich nations, which contribute to the global warming, should compensate poor ones for climate-related damages. The United Nations Framework Convention on Climate Change employs the following definition of climate change: "A change of climate which is attributed directly or indirectly to human activity that alters the composition of the global atmosphere and which is, in addition to natural climate variability, observed over comparable time period." (UNFCCC, Article 1 section 2). The issue of climate change loss and damage compensation was raised at United Nations climate talks in Warsaw in November 2013. Philippines President Benigno Aquino III has urged countries to take moral responsibility and assure that the developing world is compensated to the level that vulnerable countries and communities will become more resilient.[1] The Philippines is regarded as one of the economies that are "hot spots" for extreme weather-related disasters, which are becoming more costly and deadly. In addition, there is strong evidence of increasing temperature and more occurrences of extreme rainfall events in the Philippines in comparison to the beginning of the twentieth century.

Global warming reveals problems related to both environmental externalities and equity at the global level. Global warming is created through the "greenhouse gas (GHG) effect," a process in which gases build up a thermal blanket around the earth that traps energy from the sun at an average temperature of about 14 degrees Centigrade (57 degrees Fahrenheit). Natural disasters are one of the most visible results of global warming (Intergovernmental Panel on Climate Change 2007a, 45–54). The 4th Assessment Report of IPCC (2007b, 7) explicates that "[t]he earth's natural greenhouse effect makes life, as we know it, possible . . . human activities, primarily the burning of fossil fuels and clearing of forests have greatly intensified the natural greenhouse effect, causing global warming." Polluters are emitting countries that acted wrongfully, and thereby hold the "Historical Emission Debt (HEDi) of a country" (Neumayer 2000, 186). Neumayer justifies the liability for global warming on what "holds countries accountable for the amount of greenhouse gas emissions remaining in the atmosphere emanating from a country's historical emissions. It demands that the major emitters of the past also undertake the major emission reductions in the future as the accumulation of greenhouse gases in the atmosphere is mostly their responsibility and the absorptive capacity of nature is equally allocated to all human beings no matter when or where they live" (ibid.).

Inevitably, poor people in the least developed countries are more susceptible to environmental disaster from crop losses, storms, flooding, drought, heat-waves, or disease; are least able to cope with climate change; and therefore bear a disproportionate burden of global warming and harms arising from climate change.[2] Bangladesh provides a good example of the disproportionately harmful impact on developing countries. Bangladesh is a densely populated country; as of March 15, 2011, 142.3 million people (census 2011 result) were living in a territory of about 144,000 square kilometers. Bangladesh is located at the delta of three major rivers, and is one of the countries most vulnerable to climate change, which often results in increased natural floods, tornados, and cyclones.[3]

The likely effects of global warming in Bangladesh exemplify the unequal burden of environmental harms that result from human-induced activities that increase the level of greenhouse gases into the atmosphere (ibid.). Although the contribution of Bangladesh itself to global warming is minimal and thus has minimal responsibility for causing climate change, Bangladesh will continue to suffer from the adverse effects of global warming.

Just as poor nations often bear a disproportionate burden of global warming and other environmental degradation effects compared with wealthier nations, poor communities may bear a disproportionate burden from climate change within the same country. The floods following Hurricane Katrina in 2005 caused the deaths of nearly 1,000 people—most of them were poor residents in New Orleans (Waltham 2005). Thus, even if disaster strikes a high-income, technologically advanced nation, poor communities remain vulnerable.

The impacts of climatic changes have been addressed by the international community. Since the late 1960s, two international environmental conferences were held by the United Nations (UN): UNESCO's Biosphere Conference (1968) and the United Nations Conference on the Human Environment (1972) (Caldwell 1996). These conferences did not yield a comprehensive policy, but rather resulted in the negotiation of multilateral environmental agreements and establishment of the Man and Biosphere Program and the United Nations Environment Programme (UNEP). In 1992, the United Nations Framework Convention on Climate Change (UNFCCC) was established through negotiations at the United Nations Conference on Environment and Development (UNCED). Article 2 of the treaty states the aim to "stabilize greenhouse gas concentrations in the atmosphere at a level that would prevent dangerous anthropogenic interference with the climate system."[4] The treaty does not include an enforcement mechanism on greenhouse gas emissions for individual countries, but rather imposes non-legally binding limits on greenhouse gases. The parties to the UNFCCC convention have gathered annually since1995 in Conferences of the Parties (COP) to evaluate the risks and impacts associated with climate change. In 1997, the Kyoto Protocol was established as a legally binding set of obligations for developed countries to reduce their greenhouse gas emissions.[5] In 1998, the World Meteorological Organization (WMO) and the United Nations Environment Programme (UNEP) initiated the Intergovernmental Panel on Climate Change (IPCC). In 2002, the IPCC proposed that impacts "will fall disproportionately upon developing countries and the poor persons within all countries" (Intergovernmental Panel on Climate Change 2002, 12).

The following sections examine which theory of global distributive justice offers the fairest means of addressing the duty to bear the burdens caused by climate change, including the provision of assistance or compensation from those who are most responsible for causing climate change.

Analyzing Climate Change in International Disaster Management

What Is Distributed?

The global distributive justice implications of climate change policy depend on the ethical position taken with regard to the distribution of burdens of global climate change.[6] There are two distinct kinds of burdens of global climate change: climate adaptation and mitigation.[7] The Intergovernmental Panel on Climate Change (IPCC) defines adaptation as the "adjustment in natural or human systems to a new or changing environment. Adaptation to climate change refers to adjustment in natural or human systems in response to actual or expected climatic stimuli or their effects, which moderates harm or exploits beneficial opportunities. Various types of adaptation can be distinguished, including anticipatory and reactive adaptation, private and public adaptation, and autonomous and planned adaptation." Thus, climate refers to the capacity of a given system to adjust to climate change to reduce vulnerabilities to climate change and minimize its negative impacts. While adaptation deals with the effects of climate change, mitigation deals with the causes of climate change. The IPCC defines mitigation as: "An anthropogenic intervention to reduce the sources or enhance the sinks of greenhouse gases." Mitigation then is a proactive strategy aimed at eliminating or reducing the long-term risk of climate change to human life and property. For example, Article 12 of the Kyoto Protocol states that the "purpose of the clean development mechanism shall be to assist . . . [developing states] . . . in achieving sustainable development and in contributing to the ultimate objective of the [UN Framework Convention on Climate Change (UNFCCC)], and to assist . . . [developing states] . . . in achieving compliance with their quantified emission limitation and reduction commitments under Article 3 [of the UNFCCC]." Such mitigation efforts lead to giving credits to certain countries (those listed in Annex I[8]) for making a sacrifice that enables developing countries to pursue development projects in a way that does not emit high levels of GHGs.[9]

It should be noted that whereas mitigation efforts (that is emissions reduction) are enhanced for the most part in rich countries, adaptation efforts are addressed by both poor and rich countries. However, poor countries may have lower capacity to cope with the effects of climate change and fund partially adaptation efforts.[10]

In his book *One World*, Peter Singer discusses the issue of sharing the burdens of climate change in the chapter entitled "One Atmosphere" (Singer 2004, 14–50). Singer refers to the subject of distribution, that is, the burden of emissions into the atmosphere, as follows: "There can be no clearer illustration of the need for human beings to act globally than the issues raised by the impact of human activity on our atmosphere. That we all share the same planet came to our attention in a particularly pressing way in the 1970s when scientists discovered that the use of chlorofluorocarbons (CFCs) threatens the ozone layer shielding the surface of our planet from the full force of the sun's ultraviolet radiation" (Singer 2004, 14). Singer uses the analogy of a common sink to explain how the atmosphere has long been regarded as having infinite capacity: "Think of the atmosphere as a giant global sink into which we can pour our waste gases. Then once we have used up the capacity of the atmosphere to absorb our gases without harmful consequences, it becomes impossible to justify our usage of this asset by the claim that we are leaving 'enough and as good' for others. The atmosphere's capacity to absorb our gases has become a finite resource on which various parties have competing claims. The problem is to allocate those claims justly" (Singer 2004, 29).

Singer's theory of global distributive justice supports sharing both adaptation and mitigation burdens of climate change: "The resources the world is proposing to put into reducing greenhouse gas emissions could be better spent on increasing assistance to the world's poorest people, to help them develop economically and so cope better with climate change" (Singer 2004, 23).

According to Pogge, adaptation costs are already shared by poor countries. "The global poor get to share the burdens resulting from the degradation of our natural environment while having to watch helplessly as the affluent distribute the planet's abundant natural wealth among themselves" (Pogge 2001, 64). Adaptation is then seen as passive rather than as an active adjustment in response to climate changes since "resources are of merely instrumental significance, [and] are important only insofar as they give persons opportunities to pursue their goals . . . Like rights and access to money, so the abilities to be well nourished and to move about are of mostly instrumental importance" (Pogge 2002, 35). Hence, for Pogge the primary burden-sharing problem is to distribute funding responsibilities (negative rights) to affluent countries to fund mitigation efforts in poor countries as a proactive strategy to offer opportunities, and as such they will be regarded as intrinsically valuable. Under the scheme of the GDR, the international community can monitor the damage sustained from the activities of each state so that funds can be transferred for an appropriate purpose, be it adaptation or mitigation: "Eradicating global poverty through a scheme like the GRD also involves more realistic demands than a solution through private initiatives and conventional development aid. Continual mitigation of poverty leads to fatigue, aversion, and even contempt. It requires the more affluent citizens and governments to rally to the cause again and again, while knowing full well that most others similarly situated contribute nothing or very little, that their own contributions are legally optional and that, no matter how much they give, they could for just a little more always save yet further children from sickness or starvation" (Pogge 2001, 72).

By contrast, Sen holds that all types of capabilities are intrinsically valuable as they are coupled with freedom: "Capability is . . . the substantive freedom to achieve alternate functioning combinations ([that is, combinations of the various things a person may value doing or being] or, less formally put, the freedom to achieve various lifestyles)" (Sen 1999, 75). To the extent that we are concerned about distribution of adaptation and mitigation burdens, poor countries should think not simply about resources we have available to cope with climate change (adaptation), but their ability to transform those resources into actual capabilities (mitigation), since variations across poor countries will lead to different capabilities for the same inputs of resources. Sen's account tends to agree with Pogge for thinking about rich countries' obligations to fund mitigation efforts in poor countries; however, due to differences of capability sets, these sets are ranked differently by countries with different conceptions of what is good in life. "In dealing with responsible adults, it is more appropriate to see the claims of [the] individual on the society in terms of freedom to achieve rather than actual achievements" (Sen 1992, 148). Because of this, rich countries' commitment to pursuing mitigation is seriously limited, since for a substantial range of capabilities, there is no common basis for determining what increases these capabilities.

The Community of Justice

The distribution of burdens of global climate change raises the scope of the community of justice. Such distribution needs to identify victims who have claims against wrongdoers (polluters/payers) be it states, business corporations, or individual citizens. According to Singer: "Many of the world's poorest people, whose shares of the atmosphere's capacity have been appropriated by the industrialized nations, are not able to partake in the benefits of this increased productivity in the industrialized nations—they cannot afford to buy its products—and if rising sea levels inundate their farm lands, or cyclones destroy their homes, they will be much worse off than they would otherwise have been" (Singer 2004, 30).

Compensation is claimed by the international community, represented by some appropriate agency established for the purpose, and the funds transferred for adaptation or mitigation efforts. Such an agency monitors each state's contribution to cumulative greenhouse gas (GHG) emissions. This creates a basis for reciprocity, which constitutes the foundation of obligations between states. However, the condition of reciprocity cannot explain our environmental duties to future generations and species forming the community of humankind as a whole.[11] Specifically, it does not have the ability to solve the motivation problem relative to what morality requires on behalf of future generations while retaining considerable power to force reductions. Singer expresses his concern of focusing only on states, for states gain moral worth only through their relationship to units of moral concern, such as future generations: "The ethics we have now does go beyond a tacit understanding between beings capable of reciprocity . . . why limit morality to those who have the capacity to enter into agreements with us, if in fact there is no possibility of them ever doing so? Rather than cling to the husk of a contract view that has lost its kernel, it would be better to abandon it altogether, and consider, on the basis of universalisability, which beings ought to be included within morality" (Singer 2003, 81–82).

Pogge refers to states and in particular to supranational institutions as members of the community of justice as they have the power to regulate emissions and have taken on the international obligation to do so as they are part of the globally interdependent order. This view is entrenched in Pogge's GDR proposal: ". . . the GRD reform can produce great ecological benefits that are hard to secure in a less concerted way because of familiar collective-action problems: each society has little incentive to restrain its consumption and pollution, because the opportunity cost of such restraint falls on it alone while the costs of depletion and pollution are spread worldwide and into the future. Proceeds from the GRD are to be used toward ensuring that all human beings will be able to meet their own basic needs with dignity. The goal is not merely to improve the nutrition, medical care and sanitary conditions of the poor, but also to make it possible that they can themselves effectively defend and realize their basic interests" (Pogge 2001, 67). Indeed, interdependent states cannot realize their environmental effects on their own. This is why international institutions are needed. They provide the "rules of the game" that facilitate the management of trans-boundary environmental resources: "[a] society's basic mode of economic organization; [b] the procedures for making social choices through the conduct of, or interactions among, individuals and groups, and limitations upon such choices; [c] the more important practices regulating civil (noneconomic and nonpolitical) interactions, such as the family or the education system; [d] and the procedures for interpreting and enforcing the rules of the scheme" (Pogge 1989, 22–23).

The aim of supranational institutions is not only to enforce but also to persuade states, through the threat of future costs and damages, to reduce their emissions. As stated by Pogge: "This prudential consideration has a moral side as well. A future that is pervaded by radical inequality and hence unstable would endanger not only the security of ourselves and our progeny, but also the long-term survival of our society, values and culture. Not only that: such a future would, quite generally, endanger the security of all other human beings and their descendants as well as the survival of their societies, values and cultures. And so the interest in peace—in a future world in which different societies, values and cultures can coexist and interact peacefully—is obviously also, and importantly, a moral interest. Realizing our prudential and moral interest in a peaceful and ecologically sound future will—and here I go beyond my earlier modesty—require supranational institutions and organizations that limit the sovereignty rights of states more severely than is the current practice" (Pogge 2001, 73). The GRD proposal then stretches the boundaries of the community of justice to include future generations: "The GRD proposal is morally compelling. It can be broadly anchored in the dominant strands of western normative political thought outlined in the second section. And it also has the morally significant advantage of shifting consumption in ways that restrain global pollution and resource depletion for the benefit of future generations in particular. Because it can be backed by these four important and mutually independent moral rationales, the GRD proposal is well-positioned to benefit from the fact that moral reasons can have effects in the world" (Pogge 2001, 72).

When assessing where burdens of adaptation and mitigation to climate change should fall and which members of the community of justice owe duties, and to whom, Sen prefers to use the criteria of capabilities rather than degree of connection or reciprocity between members of the community of justice: ". . . assessment of justice demands engagement with the 'eyes of mankind,' first, because we may variously identify with the others elsewhere and not just with our local community; second, because our choices and actions may affect the lives of others far as well as near; and third, because what they see from their respective perspectives of history and geography may help us to overcome our own parochialism" (ibid., 130). The process of sharing the burdens of climate change will inevitably involve political decisions that prioritize particular national interests while vulnerable groups generally have limited input into the public management of global climate-related harms. A cosmopolitan concern for justice among individuals is seen by Sen as distinct from a concern for effective adaptation based on national preferences. Thus the goal of an international climate adaptation or mitigation funding agreement is to address injustice between individuals and communities, rather than injustice between states, so that adaptation and mitigation measures actually benefit the most vulnerable members of the community of justice. By drawing on individuals and their capabilities in a more and more restricted world, Sen excludes future generations from membership in the community of justice, "as development that prompts the capabilities of present people without compromising capabilities of future generations" (Sen 2013, 11).

Preconditions of Just Distribution

Any analysis of global distributive justice must assess the conditions that create an "entitlement" to compensation that arises from wrongdoing. The responsibility for harm justifies redistribution between people. For example, the "polluter pays" principle serves as a basis for sharing the burdens of adaptation and mitigation to climate change (Caney 2005, 767–69). The historical contribution of different countries to climate change is a crucial factor in establishing responsibilities of countries for compensation. However, it entails the need to draw the line between historical and current contributions, whereas historical emissions may have long-lasting effects on climate conditions in the future. Since the atmospheric lifetime of most greenhouse gases is long (Montenegro et al. 2007), past emissions can affect the climate in the future as a result of indirect effects related to the long lifetime of greenhouse gases (IPCCa 2007).

Despite such complexity, few attempts have been made to use historical responsibility for assessing the contributions of countries to global climate change (Olivier et al. 2014). In 1997 the Government of Brazil (the "Brazilian proposal") proposed a model that provides an inter-comparison exercise for attributing contributions to climate change[12] (UNFCCC 1998). The proposal suggested that emissions for all Annex I Parties were to be shared among individual Annex I Parties in proportion to their share of responsibility for climate change. The proposal used a simple climate model for estimating the temperature increase resulting from emissions of different countries. Although the scientific and methodological parts of the proposal were debated, the Brazilian Proposal is carried out by the UNFCCC officially.

Given the complexity inherent in attributing historical responsibility for greenhouse gases and their effects, Singer proposes using 1990, the year of the earliest assessment report established by the IPCC that first raised the issue of greenhouse gases and their effects on climate change, as the baseline. According to Singer, before 1990, people were ignorant about the effects of GHG on climate change. The condition of ignorance is vindicated by Singer: "The historical view of fairness just outlined puts a heavy burden on the developed nations. In their defense, it might be argued that at the time when the developed nations put most of their cumulative contributions of greenhouse gases into the atmosphere, they could not know of the limits to the capacity of the atmosphere to absorb those gases . . . At least since 1990, however, when the Intergovernmental Panel on Climate Change published its first report, solid evidence about the hazards associated with emissions has existed" (Singer 2004, 34).

Thus, Singer's condition of ignorance addresses wrongfulness in situations where the actor "intentionally" and "consciously" caused the harm: that he acted with the very purpose of causing it, knew that it is a by-product of his action, or at least had an idea to a certainty or near-certainty that it would result from his emissions.

Pogge also refers to the historical conditions underlying the global sharing of costs of adaptation and mitigation: "The present circumstances of the global poor are significantly shaped by a dramatic period of conquest and colonization, with severe oppression, enslavement, even genocide, through which the native institutions and cultures of four continents were destroyed or severely traumatized. This is not to say (or to deny) that affluent descendants of those who took part in these crimes bear some special restitutive responsibility toward impoverished descendants of those who were victims of these crimes. The thought is rather that we must not uphold extreme inequality in social starting positions when the allocation of these positions depends upon historical processes in which moral principles and legal rules were massively violated. A morally deeply tarnished history should not be allowed to result in radical inequality" (Pogge 2001, 65). However, both Singer and Pogge downsize the complexity of identifying and disentangling historical contributions to climate change.

Currently, the industrialized countries, such as the United States, Japan, and Russia, and many European countries, bear the chief responsibility for past contributions to climate change (den Elzen et al. 2005, Srinivasan et al. 2008). For this reason, many developing countries have claimed compensation for historical emissions made by rich countries (Najam et al. 2003). However, if assessment of responsibility of countries to share the burdens of climate change is made by referring to the conditions of historical contribution and present day inequality, one must wonder why China should be owed a historical debt by developed countries. Despite the rapid industrialization of China, China still remains a relatively poor country with a high level of inequality; thus its degree of responsibility for climate change is similar to that of developing countries (Harris 2010). For that, Sen offers to look inside the state at the distribution of wealth, emissions, and adaptation and mitigation costs, which leads to a more complex assessment of historical contributions as a condition for global distributive justice. Sen draws on the notion of functionings; that is, the actual conditions of people's doings and beings. In a society where the conditions under distributive justice ensure basic capabilities to substantive freedoms, rather than preferences shaped by an acceptance of a given order or unjust background, conditions "enhance the ability of people to help themselves and to influence the world" (Sen 1999, 18). When applied to the case of China, even that significant number of Chinese people, estimated at least in the tens of millions, are like the global rich (even wealthier than the global average) and therefore contribute to climate change as individuals in affluent countries. At the same time, China's rural poor are owed significant adaptation assistance. People's functionings and capabilities should be taken into account when assessing relative vulnerability and the responsibility of countries to share the burdens of climate change.

THE PRINCIPLE OF DISTRIBUTION (CONTENT)

One of the prevailing principles in addressing the issue of responsibility to bear the burdens of global climate change is the "polluter pays" principle (hereafter PPP). The PPP was adopted by the Organization for Economic Co-operation and Development (OECD) on May 26, 1972, and November 14, 1974. In April 2004 the European Union and Council of Ministers adopted the European Directive (2004/35/CE) on Environmental Liability with regard to the Prevention and Remedying of Environmental Damage. This directive follows the PPP: "The prevention and remedying of environmental damage should be implemented through the furtherance of the 'polluter pays' principle, as indicated in the Treaty and in line with the principle of sustainable development. The fundamental principle of this Directive should therefore be that an operator whose activity has caused the environmental damage or the imminent threat of such damage is to be held financially liable, in order to induce operators to adopt measures and develop practices to minimise the risks of environmental damage so that their exposure to financial liabilities is reduced" (Directive 2004/35/CE of the European Parliament and of the Council of 21 April 2004 on environmental liability with regard to the prevention and remedying of environmental damage, section 2).[13] The IPCC (2001) also made reference to PPP in its third assessment report entitled *Climate Change 2001: Mitigation*.

Given this, let us consider Singer's principle of justice, which is qualified with the condition of ignorance. Singer suggests the equal per capita shares (EPCS) principle, according to which every person is entitled to an equal per capita share of emission rights, compatible with maintaining an atmosphere for all people at the global level and for future generations as well. The EPCS principle addresses the responsibility of the polluters for costs of global climate change post-1990: "For at least a century the developing nations are going to have to accept lower outputs of greenhouse gases than they would have had to, if the industrialized nations had kept to an equal per capita share in the past. So by saying, 'forget about the past, let's start anew,' the pure equal per capita share principle would be more favorable to the developed countries than a historically based principle would be.... The claim that the [Kyoto] Protocol does not require developing nations to do their share does not stand up to scrutiny. Americans who think that even the Kyoto Protocol requires America to sacrifice more than it should are really demanding that the poor nations of the world commit themselves to a level that gives them, in perpetuity, lower levels of greenhouse gas production per head of population than the rich nations have. How could that principle be justified?" (Singer 2002, 44).

Thus, per capita share is calculated at about 1 metric ton per year (Singer 2002, 35). Following that, Japan and Western Europe per capita emissions range between 1.6 and 4.2, with most below 3. Developing countries average 0.6, with China at 0.76 and India at 0.29. What follows according to the equal per capita shares principle is that India and China are allowed to increase their carbon emissions up to a sustainable quota (perhaps with some concerns over population size), but the United States (U.S.) should reduce its emissions to no more than one-fiftieth of existing levels (Singer, 2002). Consequently, the EPCS principle allows countries that need a higher quota to buy it from those that emit less than their quota. It is then suggested that the U.S. could offer India not to emit its full quota, in order to be able to proceed in making a few emissions above the U.S. quota. The employment of EPCS principle then leads to consider the emission quota as a valuable asset for low-emitting countries to sell and benefit "that we support the second principle, that of equal per capita future entitlements to a share of the capacity of the atmospheric sink, tied to the current United Nations projection of population growth per country in 2050" (Singer 2002, 43).

Singer's EPCS principle may face the challenge of population size. If the population size is raised, the per capita share of emissions per year will decrease to maintain the emission rate at a sustainable level. Singer proposed that each country would be responsible for their population size: "[T]he per capita allocation could be based on an estimate of a country's likely population at some given future date. For example, estimated population size for the next 50 years, which are already compiled by the United Nations, might be used" (Singer 2002, 43). The application of Singer's EPCS principle is also problematic in the case of developed countries. People in developed countries may find it difficult to reduce their emissions since their standard of living requires more emission than in developing countries. According to the EPCS principle, developed countries have the opportunity to buy the quota of developing countries to balance their extra quota emissions.[14] This brings us to a further problem inherent to emission trading, where the corrupt government of a developing country misuses the funds from emission trading. Singer proposes then the role of international authority in emission trading: "The sale of quotas could be managed by an international authority answerable to the United Nations" (Singer 2002, 49).

Singer's EPCS principle ignores responsibility for past emissions, which is given more weight through Pogge's principle of "polluter pays," according to which states are required to compensate one another for their contributions to climate change. As seen, Pogge's principle of justice offers to alter the unjust global institutions through compensation that provides remedy for harmful or unjust behavior such as a state's historical emissions and ongoing emissions. This leads him to come up with a global redistributive mechanism, namely the GRD, to compensate the poor for their "inalienable stake in all scarce resources," which makes them more vulnerable in coping with climate change effects: "It is true that they can rent out their labor and then buy natural resources on the same terms as the affluent can. But their educational and employment opportunities are almost always so restricted that, no matter how hard they work, they can barely earn enough for their survival and certainly cannot secure anything like a proportionate share of the world's natural resources. The global poor get to share the burdens resulting from the degradation of our natural environment while having to watch helplessly as the affluent distribute the planet's abundant natural wealth among themselves. With average annual per capita income of about $82, corresponding to the purchasing power of $326 in the United States, the poorest fifth of humankind are today just about as badly off, economically, as human beings could be while still alive" (Pogge 2001, 64). The GDR will require nations to assess a dividend (tax) on any resources that they use or sell, which will result in a "tax on consumption" to alleviate poverty. However, the PPP under the scheme of GDR does not address the problem of overconsumption. In fact, it excludes those who are most vulnerable while allowing others to continue benefiting disproportionately.

Sen offers to tackle the issue of capability of polluters to pay, which is overlooked by the polluter pays and the EPCS principles. For that, the capacity to pay principle (CPP) or "beneficiary pays" principle, which ascribe duties, the most advantage follows Sen's basic capability equality principle, which maintains that each person equally has freedom to attain a sufficient level of basic capabilities. Capacity here denotes "common but differentiated responsibilities," which is articulated by the UNFCCC (UN 1995, 5). According to CPP those countries endowed with the capacity to pay are required to share the burdens of emissions. Viewed in this way, Sen does not consider all the means of well-being, such as the availability of property, social institutions, and so forth, as important and therefore cannot be the ultimate ends of well-being, but rather means to achieve higher goals such as mitigation and adaptation to global climate change: "It should be clear that we have tended to judge development by the expansion of substantive human freedoms—not just by economic growth (for example, of the gross national product), or technical progress, or social modernization. This is not to deny, in any way, that advances in the latter fields can be very important, depending on circumstances, as 'instruments' for the enhancement of human freedom. But they have to be appraised precisely in that light—in terms of their actual effectiveness in enriching the lives and liberties of people—rather than taking them to be valuable in themselves" (Drèze and Sen 2002, 3). The CPP then justifies support for adaptation and mitigation projects, or rebuilding communities affected by natural disasters. Priority will be given to the least well-off and most vulnerable communities to ensure access to funding, and reduce dependence on states that may not have the capacity to effectively or sufficiently distribute funding.

THE PRINCIPLE OF DISTRIBUTION (STRUCTURE)

Singer's EPCS principle is based on a structure laid by universalizability which "requires us to go beyond 'I' and 'you' to the universal law, the universalisable judgment, the standpoint of the impartial spectator or ideal observer, or whatever we choose to call it" (Singer 2003, 12). Universalizability is consistent with equality of obligation irrespective of nationality. According to Singer: ". . . the universalisability of ethical judgments requires us to go beyond thinking only about our own interests, and leads us to take a point of view from which we must give equal consideration to the interests of all affected by our actions. We cannot hold that ethical judgments must be universalisable and at the same time defines a person's ethical principles as whatever principles that person takes as overridingly important—for what if I take as overridingly important some non-universalisable principle like 'I ought to do whatever benefit me'? . . . Taking ethics as in some sense necessarily involving a universal point of you seems to me a more natural and less confusing way of discussing these issues" (Singer 2003, 315–16). This means that sharing the burdens of emissions is likely to be achieved through international agreements that are somewhat statist. However, if each state acts as an authority ensuring reduction of emissions within its borders, structured international institutions can implement the EPCS principle. An example for such structuring is the International Criminal Court (ICC). The ICC is designed to enforce universal punishment of crimes against humanity, while the responsibility for implementing ICC provisions is held mostly by states. Thus the supranational aspects of the ICC's jurisdiction apply only when states fail to voluntarily employ established principles.

The structure of universalizability grounded in PPP justifies compensation targeting disadvantaged people who are entitled to receive restitution for wrongful actions. Pogge is concerned with those structures that connect institutions (such as adaptation institutions in the case of climate change) with individual human rights: "[E]very human being has a global stature as the ultimate unit of moral concern" (Pogge 1992, 49). Thus, universalism built into international institutions' agreement will allow states and/or international institutions to serve as channels through which individuals will be able to fulfill their obligation to reduce emissions by taxation or regulatory measures. Prospective international agreements might lead to creating a mechanism (e.g., climate change compensation fund) that incorporates mitigation objectives in a climate adaptation fund. This structure—with a centralized fund—places the responsibility for ongoing emissions on polluters and individuals rather than states, and is more likely to ensure that compensation funds reach adaptation and mitigation projects.

Sen favors the structure of universalizability grounded in PPP, which encourages a dynamic and long-term oriented perspective in order to advance preventive actions. However, he prefers to widen the universal grounding to comprise the idea of "open impartiality," that is, "the procedure of making impartial assessments" that invokes more flexibility by permitting a wide range of agencies to apply for funding to carry out adaptation and mitigation projects, instead of only permitting states to claim compensation for climate change effects (Sen 2009, 123). Open impartiality adopted in CPP's structure allows taking into account the ". . . 'enlightenment relevance' (and not just 'membership entitlement') of views from others; secondly, the comparative (and not just transcendental) focus of Smith's investigation, going beyond the search for a perfectly just society; and thirdly, Smith's involvement with social realizations (going beyond the search only for just institutions)" (Sen 2009, 134). By allowing flexibility of the CPP principle through open impartiality in the face of uncertainty about future climate change impacts, agreement may promote the use of various measures besides financial funding to address vulnerability and impoverishment of disadvantage communities, such as education: "It should be clear that we have tended to judge development by the expansion of substantive human freedoms—not just by economic growth (for example, of the gross national product), or technical progress, or social modernization. This is not to deny, in any way, that advances in the latter fields can be very important, depending on circumstances, as 'instruments' for the enhancement of human freedom. But they have to be appraised precisely in that light—in terms of their actual effectiveness in enriching the lives and liberties of people—rather than taking them to be valuable in themselves" (Drèze and Sen 2002, 3).

Summing Up

The consequences of unmitigated climate change—such as natural disaster—pose a unique challenge to international disaster management related to global distributive justice. Climate vulnerabilities are not evenly distributed and poor countries often bear a disproportionate burden of environment degradation caused by greenhouse gas emissions. Hence, the question of who should be held responsible for damages and whether those responsible should compensate those who suffer from them is a pressing issue in international disaster management ethics.

In this chapter we have reflected on the distributive justice implications of climate change that depend on ethical positions taken with regard to which burdens and benefits of climate change are distributed, who should be held responsible and who should be compensated, the conditions justifying the allocations of responsibilities between countries, and the principles of justice applied to climate change burden sharing.

The relevant international burden sharing is adaptation and mitigation costs. While Singer supports the obligation for mitigation or adaptation as compensation for taking more than a fair share, Pogge and Sen are more concerned with ascribing obligations to affluent countries to fund mitigation efforts in poor countries as a proactive strategy to create opportunities for development. The problem of affluent countries' commitment to pursuing mitigation is raised by Sen's capabilities approach. Drawing on the capability approach could be an important complicating factor in allocation of responsibilities between countries as it forms differentiated responsibilities and respective capabilities among countries.

When assessing where burdens of adaptation and mitigation to climate change should fall and which members of the community of justice owe duties, and to whom, there is considerable agreement among the philosophers. They all assume that states remain relevant actors to hold responsibility as they have the authority to regulate emissions and have taken on international legal obligations to do so. However, Sen suspects that such an institutional aspect will end up being problematic if it cannot recognize the different contributions and capabilities of individuals and communities within states. For Sen, the goal of an international climate adaptation or mitigation funding agreement is to address injustice between individuals and communities, rather than injustice between states, so that adaptation and mitigation measures actually improve the conditions of the most vulnerable members of the community of justice.

The three philosophers also agree with the historical and current contributions of different countries to climate change as the conditions justified allocation of responsibilities for international compensation associated with adaptation and mitigation to climate change. Using universalizability as the structural grounding for principles of justice applied to climate change burden sharing, the victims of climate change effects have a right to compensation for the damage caused. Singer offers the EPCS principle, according to which every person is entitled to an equal per capita share of emission rights. Pogge supports the PPP while giving more weight to responsibility for past emissions, which obliges states to compensate one another for their contributions to climate change. Sen exposes the importance of the capacity to pay principle (CPP) or "beneficiary pays" principle that would reach inside states to target the most vulnerable people. This would entail more than economic measures but rather measures such as proper education, good governance, and so on in fostering successful adaptation and mitigation projects.

8

Code of Ethics for the International Disaster Management Practice

> The relationship between management and leadership for organizational ethics is thus both more subtle and outward . . . It links organizational purpose to the terms upon which that purpose is focused. It focuses on nurturing the relationships that sustain organizational purpose and give it its ethical character
>
> —Cox 2009, 117–18

This chapter applies the global justice framework developed in the previous chapters to evaluate current institutional arrangements set by humanitarian aid and relief organizations in international disaster management such as the Code of Conduct for the International Red Cross and Red Crescent Movement and NGOs in Disaster Response Programmes, the Humanitarian Charter and "The Sphere Project" (Sphere), the Humanitarian Accountability Partnership (HAP), "People In Aid Code of Good Practice," and the "Good Humanitarian Donorship" initiative. The next section makes the case for Codes of Ethics as institutional structures that generate organizational legitimacy. Codes of Ethics are viewed as an important management tool for promoting international disaster management to meet global justice criteria. Subsequently, the chapter sets forth a Code of Ethics that covers the ethics spectrum, touching on the major issues of concern in international disaster management and developing a comprehensive framework for effective implementation of a Code of Ethics in international disaster management in a way that adjusts to evolving challenges faced by humanitarian aid and relief organizations.

Codes of Ethics in the Service of Institutions' Legitimacy

Organizations that strive to increase their reputation as ethical enterprises need to address legitimating mechanisms. Following Suchman, legitimacy is defined as "[a] generalized perception or assumption that the actions of an entity are desirable, proper or appropriate within some socially constructed system of norms, values, beliefs and definitions" (Suchman 1995, 574). Suchman's definition of legitimacy discerns the moral value of legitimacy, that is, the conformance of an organization to social values and obligations (Deephouse and Carter 2005, Scott 1995, Suchman 1995). Such a form of legitimacy becomes normative insofar as organizations must continuously make ethical evaluations should they wish to act in ways seen to be appropriate (Suchman 1995). It is argued that ethical/moral legitimacy is derived from the efforts of international organizations to act in a more globally responsible manner. By drawing on the theoretical lens of management strategy, ethical legitimacy is engaged in restoring or establishing the global citizenship of multinational organizations to promote trust in the organization's management: "Formal ethics programs can be conceptualized as organizational control systems aimed at standardizing human behaviour" (Weaver et al. 1999, 42).

Acting in a way that is responsible to the varied stakeholders of the organization is the means by which organizations gain legitimacy (Long and Driscoll 2007, O'Donovan 2002, Warren 2003, Wartick and Cochran 1985), and Codes of Ethics are meant to articulate these responsibilities: "The most common and important way in which ethics are institutionalized is through the design and implementation [of] a corporate Code of Ethics" (Johnson and Smith 1999, 1365). In other words, Codes of Ethics serve to bridge the gap between society's expectations and perceptions of the organization's performance (Boiral 2003). Such a view suggests that legitimacy may take on a strategic form; codes of ethics are adopted because it is in the self-interest of management to adopt them.

The issue of agency self-interest becomes prominent among humanitarian aid agencies since they are deeply involved in promoting and monitoring the ethical behavior of sovereign states and individuals. It underlines the need for guiding principles that assert the priority or primary objective of humanitarian agencies to be based on the "needs of vulnerable communities and individuals" rather than any other internal consideration. Disaster management fundamentally deals with a response to disaster vulnerability and losses of people's livelihoods and assets. Thus, individuals or agencies engage in disaster response and relief actions because they are deeply committed to doing the "right thing," and therefore this field is closely tied to ethics. Such motivation is entrenched, for example, in the "Code of Conduct for The International Red Cross and Red Crescent Movement and NGOs in Disaster Relief," which was formed and granted by eight of the world's largest disaster response agencies in 1994. By February 2007 an unprecedented number of 404 national and international agencies had signed the Code, meaning that they are committed to comply with its conditions or principles. The first principle of the Code articulates the social responsibility carried out by humanitarian aid agencies, which empowers them to emerge as powerful monitors of ethical practices of other actors, namely private and public actors. "The humanitarian imperative comes first: The right to receive humanitarian assistance, and to offer it, is a fundamental humanitarian principle which should be enjoyed by all citizens of all countries. As members of the international community, we recognize our obligation to provide humanitarian assistance wherever it is needed. Hence the need for unimpeded access to affected populations is of fundamental importance in exercising that responsibility. The prime motivation of our response to disaster is to alleviate human suffering amongst those least able to withstand the stress caused by disaster. When we give humanitarian aid it is not a partisan or political act and should not be viewed as such."[1]

As seen in chapter 2, within the field of disaster emergency management there is a plethora of ethical guidelines outlined in various organizational websites (e.g., Eight Principles of Disaster Management[2]; the Republic of South Africa Disaster Management Bill[3]; The Wingspread Principles: A Community Vision for Sustainability[4] and Gujarat State Disaster Management Policy[5]; and South Asia: Livelihood Centered Approach to Disaster Management—a Policy Framework[6]). These examples support the view that the field of disaster management lacks a cohesive approach in terms of ethical principles. Two questions remain: Why is this so and what are the implications of this multiplicity.

The main reasons behind this lack of cohesive framework include: (1) Differences in fundamental values and organizational mandates, for example, the Red Cross addresses disaster assistance at the community level while the World Bank targets assistance at international and national levels, though disaster management is essential to both. (2) Differences across disciplines. Divergence between aid agencies exists because different organizations address disaster management by lying on different operational perspectives; for example, a government agency uses a strategic and a relief-based operational approach while a development agency might focus on community sustainability. (3) Aid agencies may work at different stages of the disaster management process (mitigation, preparedness, response, and recovery). Each of these stages has its own set of objectives that would result in varying practices and strategies.

Such diversity reveals that disaster management is such a multifaceted field that a coherent set of ethical guidelines is unlikely ever to be formulated. This book suggests that the field of disaster management would benefit greatly from a coherent vision of ethical principles that underlie a disaster management based on global distributive justice. Having a Code of Ethics for international disaster management would also improve communication and coordination between different organizations engaged in international disaster management efforts. In the next section, we will draw on a few examples of principles of disaster management to evaluate whether and how they correspond to distributive justice criteria.

Examples of Current Principles of Disaster Management

In an attempt to promote responsible behavior and avoid the negative consequences of unethical behavior some humanitarian aid agencies have codified their standards of ethical conduct. These organizations adopt a code of conduct or a set of principles of disaster management because it communicates to the international disaster management community that they are committed to norms of appropriate behavior, and it enhances practitioners' professional status and identity since many codes provide a mechanism for professional regulation. Publicizing the existence of a code is a way of signaling that the organization values and adheres to high ethical standards (Long and Driscoll 2007).

Despite the debate about the effectiveness or impact of ethics codes, it is the purpose of this chapter to provide a comparative overview of five initiatives aimed at strengthening the quality of humanitarian assistance and increasing professionalism in disaster management. The Code of Conduct for the International Red Cross and Red Crescent Movement and NGOs in Disaster Response Programmes, Sphere Humanitarian Charter, the Humanitarian Accountability Partnership (HAP), "People In Aid

Code of Good Practice," and the "Good Humanitarian Donorship" initiative are preeminent in the field of disaster management and, interestingly, none of these initiatives specifies enforcement mechanisms for ethical violations by their members. All of these initiatives rely on aspirational codes without specific enforcement mechanisms. In an effort to compare different codes and principles of disaster management and design a potential Code of Ethics, we will begin with a review of their summarized principles (table 8.1 on pages 126–133). This review will be followed by a comparative look at these initiatives to compare and contrast the strengths of these organizations' ethics efforts as a mechanism to create a common set of principles in disaster management.

With these examples, it is possible to offer an initial view of the codes or set of principles on ethics infrastructure typology established by the OECD (OECD/PUMA 1996, 34). Ethics infrastructure ranges across two dimensions. The first dimension is based on compliance- and integrity-based ethics approaches. Compliance-based systems focus on strict obedience to rules, training, and penalties for noncompliance (the OECD defines this as the "low road"), whilst integrity-based systems focus on what "should be achieved rather than what behavior should be avoided" (OECD 1996, 59). These systems include (1) the definition of broad aspirational values, (2) "a focus on what is achieved rather than how it was achieved," and (3) "an emphasis on encouraging good behavior rather than policing and punishing errors or bad behavior" (OECD 1996, 59; the OECD defines this as the "high road"). The second dimension is a humanitarian aid–managerialism dimension. Managerialism is a regime of governmentalizing practices based on instrumental reasoning such as measures of performance, output control, and quality of service. Humanitarian aid is a fundamental expression of core values in humanitarian aid practice such as saving people's lives, alleviating suffering, and protecting human dignity.

The International Red Cross and Red Crescent Movement and NGOs in Disaster Response Programmes is the only one that fits into a compliance-based system and humanitarian aid values as it is already in the process of establishing an ethics infrastructure through a Code of Conduct. "The Sphere Humanitarian Charter" incorporates both aspirational values such as provision of impartial assistance (section 1.1) with humanitarian aid organizational values such as maintaining human dignity and preserving life. However, the charter is deficient in setting rules of conduct autonomously, but rather makes reference to existing legal terms

Table 8.1. Existing codes and principles of disaster management.

A Sample of Codes and Principles of Disaster Management	
The Code of Conduct for the International Red Cross and Red Crescent Movement and NGOs in Disaster Response Programmes[1]	1. The humanitarian imperative comes first. 2. Aid is given regardless of the race, creed or nationality of the recipients and without adverse distinction of any kind. Aid priorities are calculated on the basis of need alone. 3. Aid will not be used to further a particular political or religious standpoint. 4. We shall endeavour not to act as instruments of government foreign policy. 5. We shall respect culture and custom. 6. We shall attempt to build disaster response on local capacities. 7. Ways shall be found to involve programme beneficiaries in the management of relief aid. 8. Relief aid must strive to reduce future vulnerabilities to disaster as well as meeting basic needs. 9. We hold ourselves accountable to both those we seek to assist and those from whom we accept resources. 10. In our information, publicity and advertising activities, we shall recognize disaster victims as dignified human beings, not hopeless objects.
Sphere Humanitarian Charter and Minimum Standards in Disaster Response[2]	We reaffirm our belief in the humanitarian imperative and its primacy. By this we mean the belief that all possible steps should be taken to prevent or alleviate human suffering arising out of conflict or calamity, and that civilians so affected have a right to protection and assistance. It is on the basis of this belief, reflected in international humanitarian law and based on the principle of humanity, that we offer our services as humanitarian agencies. We will act

1. http://www.ifrc.org/en/publications-and-reports/code-of-conduct/#sthash.CJeOwRIh.dpuf
2. http://www.sphereproject.org/handbook/index.htm

in accordance with the principles of humanity and impartiality, and with the other principles set out in the Code of Conduct for the International Red Cross and Red Crescent Movement and Non-Governmental Organisations (NGOs) in Disaster Relief (1994).

1.1 The right to life with dignity

This right is reflected in the legal measures concerning the right to life, to an adequate standard of living and to freedom from cruel, inhuman or degrading treatment or punishment. We understand an individual's right to life to entail the right to have steps taken to preserve life where it is threatened, and a corresponding duty on others to take such steps. Implicit in this is the duty not to withhold or frustrate the provision of life-saving assistance. In addition, international humanitarian law makes specific provision for assistance to civilian populations during conflict, obliging states and other parties to agree to the provision of humanitarian and impartial assistance when the civilian population lacks essential supplies.

1.2 The distinction between combatants and non-combatants

This is the distinction which underpins the 1949 Geneva Conventions and their Additional Protocols of 1977. This fundamental principle has been increasingly eroded, as reflected in the enormously increased proportion of civilian casualties during the second half of the twentieth century. That internal conflict is often referred to as 'civil war' must not blind us to the need to distinguish between those actively engaged in hostilities, and civilians and others (including the sick, wounded and prisoners) who play no direct part. Non-combatants are protected under international humanitarian law and are entitled to immunity from attack.

	1.3 The principle of non-refoulement This is the principle that no refugee shall be sent (back) to a country in which his or her life or freedom would be threatened on account of race, religion, nationality, membership of a particular social group or political opinion; or where there are substantial grounds for believing that s/he would be in danger of being subjected to torture.
Humanitarian Accountability Partnership (HAP)[3]	**Humanity**: concern for human welfare and respect for the individual. **Impartiality**: providing humanitarian assistance in proportion to need, and giving priority to the most urgent needs, without discrimination (including that based upon gender, age, race, disability, ethnic background, nationality or political, religious, cultural or organisational affiliation). **Neutrality**: aiming only to meet human needs and refraining from taking sides in hostilities or giving material or political support to parties to an armed conflict. **Independence**: acting only under the authority of the organisation's governing body and in line with the organisation's purpose. **Participation and informed consent**: listening and responding to feedback from crisis-affected people when planning, implementing, monitoring and evaluating programmes, and making sure that crisis-affected people understand and agree with the proposed humanitarian action and are aware of its implications. **Duty of care**: meeting recognised minimum standards for the well-being of crisis-affected people, and paying proper attention to their safety and the safety of staff. **Witness**: reporting when the actions of others have a negative effect on the well-being of people in need of humanitarian assistance or protection.

---

3. http://www.hapinternational.org/en/page.php?IDpage=3&IDcat=10

	Offer redress: enabling crisis-affected people and staff to raise complaints, and responding with appropriate action. **Transparency:** being honest and open in communications and sharing relevant information, in an appropriate form, with crisis-affected people and other stakeholders. **Complementarity:** working as a responsible member of the aid community, co-ordinating with others to promote accountability to, and coherence for, crisis-affected people.
The "People In Aid Code of Good Practice"[4]	Human Resources Strategy *Human resources are an integral part of our strategic and operational plans*
	Staff Policies and Practices *Our human resources policies aim to be effective, fair and transparent*
	Managing People *Good support, management and leadership of our staff is key to our effectiveness*
	Consultation and Communication *Dialogue with staff on matters likely to affect their employment enhances the quality and effectiveness of our policies and practices*
	Recruitment and Selection *Our policies and practices aim to attract and select a diverse workforce with the skills and capabilities to fulfil our requirements*
	Learning, Training and Development *Learning, training and staff development are promoted throughout the organisation*
	Health, Safety and Security *The security, good health and safety of our staff are a prime responsibility of our organisation*
The "Good Humanitarian Donorship" initiative[5]	**Objectives and definition of humanitarian action** 1. The objectives of humanitarian action are to save lives, alleviate suffering and

---

4. http://www.peopleinaid.org/pool/files/code/code-en.pdf
5. http://www.goodhumanitariandonorship.org/

The "Good Humanitarian Donorship" initiative *(continued)*	maintain human dignity during and in the aftermath of man-made crises and natural disasters, as well as to prevent and strengthen preparedness for the occurrence of such situations. 2. Humanitarian action should be guided by the humanitarian principles of *humanity*, meaning the centrality of saving human lives and alleviating suffering wherever it is found; *impartiality*, meaning the implementation of actions solely on the basis of need, without discrimination between or within affected populations; *neutrality*, meaning that humanitarian action must not favour any side in an armed conflict or other dispute where such action is carried out; and *independence*, meaning the autonomy of humanitarian objectives from the political, economic, military or other objectives that any actor may hold with regard to areas where humanitarian action is being implemented. 3. Humanitarian action includes the protection of civilians and those no longer taking part in hostilities, and the provision of food, water and sanitation, shelter, health services and other items of assistance, undertaken for the benefit of affected people and to facilitate the return to normal lives and livelihoods. **General principles** 4. Respect and promote the implementation of international humanitarian law, refugee law and human rights. 5. While reaffirming the primary responsibility of states for the victims of humanitarian emergencies within their own borders, strive to ensure flexible and timely funding, on the basis of the collective obligation of striving to meet humanitarian needs.

6. Allocate humanitarian funding in proportion to needs and on the basis of needs assessments.
7. Request implementing humanitarian organisations to ensure, to the greatest possible extent, adequate involvement of beneficiaries in the design, implementation, monitoring and evaluation of humanitarian response.
8. Strengthen the capacity of affected countries and local communities to prevent, prepare for, mitigate and respond to humanitarian crises, with the goal of ensuring that governments and local communities are better able to meet their responsibilities and co-ordinate effectively with humanitarian partners.
9. Provide humanitarian assistance in ways that are supportive of recovery and long-term development, striving to ensure support, where appropriate, to the maintenance and return of sustainable livelihoods and transitions from humanitarian relief to recovery and development activities.
10. Support and promote the central and unique role of the United Nations in providing leadership and co-ordination of international humanitarian action, the special role of the International Committee of the Red Cross, and the vital role of the United Nations, the International Red Cross and Red Crescent Movement and non-governmental organisations in implementing humanitarian action

Good practices in donor financing, management and accountability
(a) Funding
11. Strive to ensure that funding of humanitarian action in new crises does not adversely affect the meeting of needs in ongoing crises.

The "Good Humanitarian Donorship" initiative *(continued)*	12. Recognising the necessity of dynamic and flexible response to changing needs in humanitarian crises, strive to ensure predictability and flexibility in funding to United Nations agencies, funds and programmes and to other key humanitarian organisations 13. While stressing the importance of transparent and strategic priority-setting and financial planning by implementing organisations, explore the possibility of reducing, or enhancing the flexibility of, earmarking, and of introducing longer-term funding arrangements. 14. Contribute responsibly, and on the basis of burden-sharing, to United Nations Consolidated Inter-Agency Appeals and to International Red Cross and Red Crescent Movement appeals, and actively support the formulation of Common Humanitarian Action Plans (CHAP) as the primary instrument for strategic planning, prioritisation and co-ordination in complex emergencies. *(b) Promoting standards and enhancing implementation* 15. Request that implementing humanitarian organisations fully adhere to good practice and are committed to promoting accountability, efficiency and effectiveness in implementing humanitarian action. 16. Promote the use of Inter-Agency Standing Committee guidelines and principles on humanitarian activities, the Guiding Principles on Internal Displacement and the 1994 Code of Conduct for the International Red Cross and Red Crescent Movement and Non-Governmental Organisations (NGOs) in Disaster Relief. 17. Maintain readiness to offer support to the implementation of humanitarian action, including the facilitation of safe humanitarian access.

18. Support mechanisms for contingency planning by humanitarian organisations, including, as appropriate, allocation of funding, to strengthen capacities for response.
19. Affirm the primary position of civilian organisations in implementing humanitarian action, particularly in areas affected by armed conflict. In situations where military capacity and assets are used to support the implementation of humanitarian action, ensure that such use is in conformity with international humanitarian law and humanitarian principles, and recognises the leading role of humanitarian organisations.
20. Support the implementation of the 1994 Guidelines on the Use of Military and Civil Defence Assets in Disaster Relief and the 2003 Guidelines on the Use of Military and Civil Defence Assets to Support United Nations Humanitarian Activities in Complex Emergencies.

(c) Learning and accountability
21. Support learning and accountability initiatives for the effective and efficient implementation of humanitarian action.
22. Encourage regular evaluations of international responses to humanitarian crises, including assessments of donor performance.
23. Ensure a high degree of accuracy, timeliness, and transparency in donor reporting on official humanitarian assistance spending, and encourage the development of standardised formats for such reporting.

and principles included in the IFRC Code of Conduct and international humanitarian law. The Humanitarian Accountability Partnership (HAP), the "People In Aid Code of Good Practice," and the "Good Humanitarian Donorship" stand at the extreme of two axes, having an integrity approach aligned with managerialism values. It concentrates on aspirational values such as humanity, neutrality, reliability (trust), and duty of care with an aim to increase institutional responsiveness and transparency in implementing humanitarian action. Such ethics infrastructure allows for accountability, transparency, and individual moral agencies to flourish. However, the risk this poses is reinforced by the extensive devolution of uniqueness of humanitarian aid functions. None of these initiatives fits into either of the two dimensions: compliance-based system and managerialism.

Table 8.1 summarizes the comparison of the existing ethics infrastructures along two dimensions—an integrity-compliance dimension and a humanitarian aid–managerialism dimension.

Most of the initiatives have introduced a value declaration (set of principles) or a code of conduct (see table 8.1). Value declarations are used to state the core values, but they usually do not provide specific guidelines on how to apply these values in practical situations. For example, the Humanitarian Accountability Partnership (HAP) uses value declaration that generally states that transparency is a core value, but value declaration does not provide guidelines on, for instance, how humanitarian aid employees can act toward vulnerable communities on projects or programs that are focused on the groundwork (e.g., development aid). Guidelines or detailed standards of behavior are set up by code of conduct. A code of conduct can be seen as an extension of value declaration that transforms the values into practice.

Both value declaration and code of conduct can be seen as two initial stages in the development of Codes of Ethics. As a first stage, organizations often begin by articulating their core values and advance them by announcing a declaration of values. As deliberation and discussion on ethics advance, the organization is ready to introduce more systematic and detailed guidelines in the form of a code of conduct. However, both value declaration and code of conduct list the organization's core values while failing to guide members of the organization in how to apply the core values in practice. Thus, the proposed Code of Ethics is based on a global distributive justice framework whose tenets refer to ethical issues faced by the international disaster management community.

Code of Ethics for International Disaster Management

Purpose

The purpose of the Code of Ethics for International Disaster Management is to:

- Promote global distributive justice by focusing on new challenges that are related to international disaster management.

- Provide an updated ethical framework that builds on cosmopolitan values, and emphasizes the implications of decision making for disaster aid agencies that is broader and more diverse than those in the past.

- Illustrate the complexity of moral reasoning that is required to maintain adherence to ethical principles, when ethical dilemmas lead to obligations conflicts between deeply held value systems or when ethical uncertainties occur.

- Demonstrate increased awareness of ethical issues involved in short- and long-term international disaster management activities as a way of helping aid agencies gain a meaningful appreciation of the big picture.

- Identify strategies for dealing with these ethical issues as they arise

Guiding Framework

This Code is organized into four major ethical dilemmas. International disaster management community members are encouraged to use the principles as a guide when reflecting on the degree to which their practice upholds those values. Each set of principles relating to an ethical dilemma is based on the multidimensionality of distributive justice judgments, including distributive good, community of justice, preconditions of distribution, the structure of principles and rules, and their content.

Code of Ethics

Ethical Dilemma: Dependency Syndrome

International humanitarian aid agencies should distribute relief assistance to meet immediate subsistence needs of a disaster-affected population in situations of acute risk to survival, including medical care, water and sanitation, food, and shelter.

International relief efforts enable disaster-affected populations to maintain their livelihood, preventing them from sliding into destitution.

Distribution of external aid should respect fundamental freedoms and autonomy to enable victims to exercise their capability to change and address their own subsistence needs.

External aid in the face of adversity should respond to subsistence needs as a result of disaster. As part of disaster response funding, donors may support disaster preparedness or disaster risk programs to strengthen the affected government's capacity to effectively respond to disasters.

In times of disaster, international aid agencies should build disaster vulnerability assessment of the affected population for targeting aid distribution without any form of discrimination on the basis of race, culture, ethnicity, gender, age, political ideology, religion, or mental/physical disability.

In making disaster vulnerability assessment for targeting aid distribution, international disaster relief agencies need to gain meaningful knowledge of local communities and assist affected communities to articulate their needs and cultural and ethnic characteristics in a way that enables relief agencies to act on them.

International disaster relief agencies should invest in transparent and reliable information of what recipients are entitled to and how assistance is likely to be provided to avoid developing a mentality of dependency syndrome among recipients.

Disaster vulnerability assessment by the international disaster management community must include assessments of donor performance, channels of cash delivery, and actual funding amounts.

Tackling dependency in international disasters, managements should operate under principles of equitable distribution of subsistence resources and services to meet immediate needs arising from disaster while supporting local communities and NGOs to exercise their capability for establishing an informal community support system in the face of a disaster event.

The structure of global interdependency enables fostering members of the global community to fulfill their civic responsibility, such as participation in global affairs and to act where possible to promote cosmopolitan justice.

Cooperative forms of disaster relief aid delivery with greater respect for disaster-vulnerable people's needs assessment ought to encourage people who are dependent on external assistance to actively participate in disaster management rather than conceiving of themselves as passive and vulnerable recipients.

Ethical Issue: Donor Fatigue

Aid agencies and donor states should be required to justify failure to act on an updated assessment of costs and benefits of disaster assistance.

Development of regulations and institutional arrangements should launch a pre-commitment device to confront future disaster to halt the effects of psychic numbing on a disaster-affected population.

Mobilization of global public sentiment is necessary to overcome donation fatigue through new media to create effective imagery using visual display of distant suffering people.

Empowering affected individuals to employ their institutional capacities (e.g., coordination) and power independently during response and relief efforts to avoid long-term beneficiary dependence on external aid.

Aid agencies should provide reliable and clear information of relief assistance and entitlements while aid recipients must ensure transparency in reporting on aid distribution and spending.

Aid agencies should guard against their own biases and values affecting vulnerability assessment and the provision of disaster aid, thus seeking to eliminate prejudicial attitudes concerning the political culture and competencies of the recipient country.

Ethical Issue: Corruption

Perceptions of corruption in a recipient country should not affect motivation to donate by international aid organizations by pointing at the possible consequences such as donor fatigue.

Fighting corruption in developing countries should address transparency and accountability of aid delivery and distribution established by collaboration between anticorruption institutions, donor, and international financial institutions as well as local governments and agencies involved in disaster response and recovery efforts.

Providing opportunities for community resilience by supporting disaster vulnerable communities' entrepreneurial projects or those who prove to be able to take alternative employment and have the capacity to adapt, especially in the face of adversity.

Enhancing collaboration between multilateral and bilateral development institutions, civil society, the private sector, and other actors for joint initiatives to mitigate corruption in disaster management.

Establishing disaster management audits and complaints channels staffed with local and international representatives necessary to strengthen disaster and post-disaster good governance and transparency.

Stressing the role of informal institutions (e.g., norms of reciprocity, reputation, third-party sanctioning) at the local level that can lead to raise awareness, develop new and renewed competencies, and build mutual support and collaborative efforts among affected community members.

Strengthening community resilience in corrupt settings must ensure a place or public facility to evacuate to at the outbreak of disaster event, having permanent shelter and maintaining good path and roads for transport.

Ethical Issue: Compensation

The relevant international burden sharing is disaster adaptation and mitigation costs.

The historical and current contributions of different countries to climate change justify allocation of responsibilities for international compensation associated with adaptation and mitigation to climate change.

Affluent countries are obligated to fund disaster mitigation efforts in poor countries as a proactive strategy to create opportunities for development.

States are considered relevant actors to hold responsibility as they have the authority to regulate emissions and have taken on the international legal obligations to do so.

Victims of climate change effects have a right to compensation for the damage caused through measures such as financial benefits, proper education, and good governance in fostering successful adaptation and mitigation projects.

Emerging Best Practice in International Disaster Management

Developing a Code of Ethics, while a valuable task in and of itself, is really a milestone of an overall process for meeting the requirements for the effective implementation of organization ethical standards. Best practice is a well-defined method for ensuring that the code and its supporting strategies continue to reflect the organization's ethical priorities. Key strategies in maintaining the effectiveness of implementation of a Code of Ethics include communication, training, enforceability, leadership, and ethical climate (Ireni Saban 2014).[7]

Communication Strategy

Best practice requires communication strategies to increase the awareness of the public of the existence of the Code. Effective communication entails publication and dissemination to the organization's staff members, existing and potential volunteers, donors, recipients (disaster-affected communities), governments, and the global community as a whole. Posting a series of scenarios of possible ethical dilemmas on the organization's website may exercise employees' ethical judgment. The international initiative of Active Learning Network for Accountability and Performance in Humanitarian Action (ALNAP) was established in 1997 to provide a sector-wide membership forum that aims for awareness and communication of ethical standards to improve performance of humanitarian aid agencies.[8] ALNAP is engaged in a range of activities including the publication of an annual *Review of Humanitarian Action*, which monitors the performance of humanitarian action though a combination of evaluative reports provided by the Membership. The ALNAP Secretariat creates a comprehensive evaluative reports database. The *ALNAP quality pro forma* works closely with aid agencies to improve their evaluation skills. ALNAP's biannual meetings provide the membership with extensive opportunities for networking and communication across aid agencies and their staff members.

Training Strategy

Best practice requires that effective implementation of a Code should seek to develop organization members' sensitivity toward ethical issues and dilemmas. The training strategy must ensure that members have the opportunity to engage in orientation and training programs such as role-playing, simulations, and other interactive activities to reinforce understanding of the serious impact of unethical behavior and the importance of reporting unethical situations. After completing the training program, organization members would be given a certificate as proof that in practice they apply the Code's standards. In addition, staff members are required to complete a training course in ethical decision making and on the Codes of Ethics, which will function as a crucial factor in determining the employee's entitlement to performance-based remuneration. Both ALNAP and Quality COMPAS undertake evaluation and training in humanitarian aid action and standards. The ALNAP initiates guidance booklets and training modules on ethics themes and other issues prioritized by the Membership. The training modules draw on lessons learned for particular types of emergencies. In 1999, Group URD and partners created an applied research project, the Quality Project. This project resulted in 2004 with the introduction of the first version of the Quality COMPAS. The Quality COMPAS is a quality assertion method including tools of training modules and consultation services, which have been designed specifically for aid agencies with the overall aim of improving aid actions in the face of an emergency event. The COMPAS training modules aim to ensure that participants gain the skills required for quality management. Group URD offers a complete training course (introduction to quality, advanced module on project quality management) and a training module for trainers in order to give humanitarian actors a broad understanding of the issues involved in humanitarian aid actions.[9]

Ethics Education Strategy

Best practice should provide an ethics education framework for staff members to be educated on professional and ethical conduct. In systems with great complexity such as international aid organizations, an ethics education program should be integrated into professional curricula at all levels as a basis for teaching the Code's content. Ethics education programs include written assignments, and a project or practicum to develop skills and competence in professional ethics and ethical decision making. It is suggested that the use of interactive e-learning tools can guide staff members on how to consider their ethical commitment as professional qualified employees according to the organization's Codes of Ethics. At later stages, an organization must develop a consultative process that will enable all staff members to comment on the effectiveness of the Codes of Ethics as a tool for resolving ethical dilemmas that they might encounter.

Enforceability Strategy

Best practice requires infrastructure to support the enforcement of codes inside and outside of the organization. Compliance with the Code of

Ethics is monitored by Internal Audit to ensure the code implementation process is appropriate to investigate alleged breaches of the Code. The ethics committee should be advisory in purpose. The function of the ethics committee should be to assist in resolving immediate and complicated ethical dilemmas and to submit their recommendations in a timely and prompt fashion in accordance with the demands of the situation and the issues involved. "People In Aid," for example, uses a number of guiding materials including a human resource (HR) Audit Toolkit, HR Manuals, and a tailored Employee Survey to evaluate staff members' satisfaction levels with organizational performance and to increase communication by focusing on wider benefits of the Code to staff members.[10]

Leadership Strategy

The commitment to the Code's guidelines must be demonstrated by ethical leadership. The development of the Code of Ethics should be endorsed by proactive agencies who seek to develop the global community for outward facing good, allowing for necessary infrastructure to ensure that accountability, transparency, and individual moral competency can be enhanced. Leadership strategies incorporate open discussions about ethics by sharing ethical problems that may have come up during disaster response and relief assistance or debated issues on whom to inform about misconduct or unethical behavior. Leaders may address the principles and the effectiveness of the Codes of Ethics regularly in their press releases, disaster management reports, meetings, speeches, and presentations—disseminating and publicizing agencies that demonstrated ethical performance through workshops and briefings.

Ethical Climate and Culture Strategy

Best practice requires that effective implementation of a Code of Ethics must be supported by an environment that cultivates it. The extent to which ethical principles become embedded in an organizational culture will be important in determining the success of the implementation process. Organizational culture represents the visible face of organizational commitment to ethical conduct through the formation of Codes of Ethics or ethical principles, while ethical climate evolves through a collection of general perceptions and attitudes that directs agreements among members of the organization on ethical decision making and practices. Ethical climate and culture strategies are pursued by informing the organization's stakeholders and target public of the presence of the Code of Ethics and the organization's staff members' commitment to ethical conduct as part of their professional performance. The organization must manage and reward a culture of accountability and transparency within the organization that benefits the general public and strengthens collaborations with other actors engaged in disaster response and relief efforts. It is suggested that development of an annual survey of the extent to which the Codes have influenced the ethical culture of the organization is a valuable tool to improve ethical decision making and conduct. By initiating an easy and safe channel for reporting violations of the Code of Ethics, the organization can encourage members to report an alleged unethical or illegal activity occurring during response and relief efforts without fear of retaliation or social isolation. These proposed actions must then demonstrate the benefits of an ethical conduct to organization members and to the global community interest as a whole.

A pioneering collaborative effort was initiated by HAP International, "People In Aid," and The Sphere Project that is called the Joint Standards Initiative (JSI). JSI aims at improving the quality and accountability of humanitarian actions. The role of the JSI Stakeholder Consultation is to supply evidence from a wide range of actors engaged in humanitarian aid efforts in order to create greater coherence for users of ethical standards of humanitarian aid during a disaster event. The JSI Stakeholder Consultation uses one-to-one interviews and focus group discussions across the globe as well as an online survey that involved over 2,000 people from 350 organizations in 114 countries to enhance consultative, open, and evidence-based response to aid-workers' needs.[11] These organizations demonstrate that most have not yet chosen to pursue the development of enforcement and training mechanisms for their Codes of Ethics. It seems that the "Good Humanitarian Donorship" initiative has become more proactive in employing an enforcement mechanism among those engaging in international aid.

Finally, Codes of Ethics are necessary but not sufficient of themselves in the pursuit of good governance. They must form part of a wider ethical framework that is itself part of a best practice framework designed for the international disaster management community. Existing initiatives to improve implementation of ethical standards and principles in humanitarian aid have not set the standards and the contents of ethics education and leadership programs. Since humanitarian aid has attained the status of a profession at the global level, ethics training and education should be considered a salient factor affecting effective disaster management. Online forums and professional networks may serve as valuable consultant and training mechanisms for aid workers faced with ethical dilemmas that need to be addressed in a timely manner. Such initiative requires collaborative efforts of various aid agencies' professional organizations in the difficult and ongoing task of defining and building mechanisms for developing "ethical competence" when faced with competing interests, loyalties, and values among their members.

Conclusion

Today, the increasing frequency and magnitude of natural disasters put unprecedented demands on international humanitarian assistance. It is being pushed to do more with less because of the global economic downturn. Today's international disaster management environment, with its shortage of staff and resources as well as economic constraints, can be overwhelming for international aid organizations who try to uphold their aid duties and ethical and professional principles. At the same time, the impact of international aid organizations' distributive decisions is expansive and influential, as a poor decision made by a humanitarian aid organization engaged in disaster relief and response efforts can seriously impact the lives of people in the disaster-affected country. The need for leadership in international disaster management is as clear today as it ever was. Leaders are assumed to have a long-term perspective on their allocative decision making. They do so by nurturing the relationships that sustain organizational managerial purpose and give it its ethical inspirational vision. As Northouse (2001) spells out in his book on leadership: "Ethics is central to leadership because of the nature of the process of influence, the need to engage followers to accomplish mutual goals, and the impact leaders have on establishing the organization's values" (255).

A central challenge in contemporary international disaster management is the need to develop and maintain leadership. Caught between their short-term goals to quickly mobilize response and relief resources and services and their longer-term aspirations in a way that achieves an equitable allocation of aid resources for all those who could be affected by the organization and its allocative decisions, the international disaster management community is grappling with this challenge in a variety of ways. Comparatively, little work has been invested in the kind of challenge addressed in the present book. This book suggests that a key element in this challenge is the cultivation of ethical leadership, especially by the application of distributive justice.

Since ethical leadership in disaster management means the capacity and the courage to make hard allocative choices and to explain and justify those choices, its practice must be combined with a distributive justice framework. In moving toward a more strategic position of disaster management for the future and by facilitating aligned and coordinated response and relief efforts, universal duty to assist should be based on the rules and norms that shape the distribution of aid resources and services by the international humanitarian aid system as a whole. Despite the fact that international aid organizations may claim that their values are timeless and universal, they are nevertheless bound by time and place. Hence, such positioning and alignment necessarily involve the codification of values and principles as well as underlying organizational objectives. Thus, the purpose of this book was twofold: first to examine through an institutional lens ethical dilemmas related to distributive justice that come to the fore, and then to assess a number of theories of global distributive justice to provide guidelines (Codes of Ethics) for dealing with them that are relevant to international disaster management. A code of ethics is considered an important management tool for building the ethical culture of a profession or organization by improving the profession's reputation and developing a deep sense of commitment to ethical conduct and pride among professional community members.

Translating general ethical principles and values into action (Codes of Ethics) entails the identification of ethical dilemmas associated with disaster aid resource allocation. This book raised four vexing ethical dilemmas: the dependency syndrome, donation fatigue, corruption, and climate change compensation. The analysis of ethical dilemmas faced by the international disaster management community in this book is based on five major facets (conceptual classifications) of distributive justice judgments: distributive good, community of justice, preconditions of distribution, the structure of principles and rules, and their content. It should be noted that each ethical dilemma is presented in outline only to indicate the lines along which each kind of evaluation should proceed. We expect the reader to carry the analysis forward to completion.

This book contributes to the debate on global distributive justice issues raised in international disaster management by providing an analysis of the four ethical dilemmas under three formalized theories of global distributive justice. The global just philosophies of Singer, Pogge, and Sen were selected for this analysis because they represent different schools of thought on global justice while sharing the demand for equitably sharing of resources and access to opportunities of welfare as components in achieving the global objectives of distributive justice at the global level.

In discussing the dilemmas of dependency syndrome, all three philosophers agree that dependence on international relief efforts is justified since it enables people to maintain their livelihood, preventing them from sliding into destitution in situations of acute risk to survival such as in disaster events. While Singer and Pogge agree that affluent countries and their citizens hold moral responsibility to alleviate suffering and death, Sen wishes affected countries and their citizens to assume such responsibility; that is, relief assistance should not be coerced or developed by someone else. Global dialogue and democratic deliberation help to lessen the tension between disaster relief and development objectives. Pogge accepts that relief distribution may lead to stigmatization of those receiving the assistance as passive and dependent, but insists on institutional mechanisms that may enforce international disaster relief agencies to invest in transparent and reliable information of what recipients are entitled to and how assistance is likely to be provided to avoid developing a mentality of dependency syndrome among recipients.

The dilemma of donation fatigue highlights the way donors' interests and priorities have dominance over recipients. Both Pogge and Sen insist that a global regime for sustainable aid flows must ensure that emerging donor governments, private companies, and NGOs are all committed to standards that multilateral actors establish. Here one finds a sharp and growing disagreement about the means to maintain aid flow among Sen and Pogge and Singer. Singer emphasizes the need to spark emotion or a feeling of suffering and passive victimization and thus motivate action and raise more donations. Singer's conviction is supported by psychological research that has long stressed that people are much more willing to aid identified individuals than unidentified or statistical victims.

Corruption in recipient countries may lead to withdrawal of support for international humanitarian aid. Both Singer and Pogge insist on an institutional arrangement to fight corruption in developing countries by improving the transparency and accountability of disaster relief aid delivery and distribution established by impartial institutions in the complex web of interdependencies at the global level. However, Sen doubts the capacity of impartial institutions alone to fight corruption in disaster aid delivery as supported by both Pogge and Sen. For that, he draws on the role of informal institutions at the local level to strengthen local community resilience in corrupt settings. Sen offers that international programs and projects should enhance the capacity of communities at risk to cope and overcome adversity or a disaster event. Community resilience targeted to harness notable strengths of "vulnerable communities" derives significant momentum for positive and optimistic change, such as communicative, educational, and inclusive coordination with nongovernmental entities, namely NGOs, local leaders, and community-based local service. For Sen, identifying and building on the local competencies of at-risk communities can promote their self-realization of capabilities and can stimulate enduring positive changes in terms of empowerment.

When assessing where burdens of adaptation and mitigation to climate change should fall and which members of community of justice owe duties, and to whom, there is considerable agreement among the philosophers. They all assume that states remain relevant actors to hold responsibility because they have the authority to regulate emissions and have taken on international legal obligations to do so. The relevant international burden sharing is adaptation and mitigation costs. While Singer supports the obligation for mitigation or adaptation as compensation for taking more than a fair share, Pogge and Sen are more concerned with ascribing obligations to affluent countries to fund mitigation efforts in poor countries as a proactive strategy to create opportunities for development. The problem of affluent countries' commitment to pursuing mitigation is raised by Sen's capabilities approach. Despite the shared use of universalizability as the structural grounding for principles of justice applied to climate change burden sharing, each philosopher applies a different principle for equitable allocation of burdens. Singer offers the EPCS principle, according to which every person is entitled to an equal per capita share of emission rights. Pogge supports the PPP while giving more weight to responsibility for past emissions, which obliges states to compensate one another for their contributions to climate change. Sen exposes the importance of the capacity to pay principle (CPP) or "beneficiary pays" principle, which would reach inside states to target the most vulnerable people. This would entail more than economic measures but rather measures such as proper education, good governance, and so on in fostering successful adaptation and mitigation projects. Thus, the application of Sen's principle has to go hand-in-hand with more investment in local capacity and resilience—which should be based, as much as possible, on a partnership among international aid organizations and government, civil society, and the private sector engaged in disaster response and relief efforts.

As these dilemmas suggest, the international disaster management community is continually confronted with value-laden choices, and the distributive questions of who gets what and how should be made explicit. This book formulates a Code of Ethics to set out the aspirational standards of behavior expected of international disaster management community members and serves the global community to meet their expectations toward the profession. The ethical standards set up by a Code for the international disaster management community, when effectively communicated to its professional community, is likely to contribute toward developing professional identity and growth that can lead to a more supportive and regulatory environment and increased public trust. Such a move toward ethics codification in international disaster management is a critical process in holding disaster response and relief practitioners accountable for compliance. This chapter suggests that the professional standards and priorities must be clearly communicated to engender international aid organizations' commitment to meeting the standards contained in the Codes. Communication of the Code may strengthen professional identity among disaster response and relief practitioners and increase confidence and comfort in making ethical decisions based on clear understanding of their moral obligations and responsibilities to the affected population as well as the global community as part of their professional integrity. Yet, the existence of a Code of Ethics does not of itself increase the likelihood of ethical behavior. Much depends on how ethical guidelines are developed, perceived, and integrated into disaster management practices if we are to ensure that humanitarian aid continues to meet the requirements and expectations of those in need. It is a time of constant challenge, but also of great opportunity.

Notes

Introduction

1. Although it is agreed by all that disaster caused by nature has myriad human causes, the forces that exact natural disasters are considered as natural phenomena that occur regardless of the presence of man.
2. InterAgency Standing Committee, Operational Guidelines on Human Rights and Natural Disasters. Washington, DC: Brookings-Bern Project on Internal Displacement, June 2006.
3. http://emdat.be/advanced_search/index.html.
4. Ibid.
5. http://earthobservatory.nasa.gov/NaturalHazards/view.php?id=14243.
6. http://ec.europa.eu/echo/files/funding/decisions/2005/dec_guyana_01000_en.pdf.
7. http://www.who.int/hac/crises/irn/Fat%20report%20Zarand%20final.pdf.
8. http://foundationcenter.org/gainknowledge/research/pdf/katrina_report_2006.pdf.

Chapter 1

1. This typology is based on IFRC classification of hazards. See https://www.ifrc.org/en/what-we-do/disaster-management/about-disasters/definition-of-hazard/.
2. http://www.oxfam.org/sites/www.oxfam.org/files/forecasting-disasters-2015.pdf.
3. EM-DAT: The CRED/OFDA International Disaster Database at http://www.emdat.be/database.
4. EM-DAT: The CRED/OFDA International Disaster Database at http://www.emdat.be/database.
5. http://www.unisdr.org/files/7817_7819isdrterminology11.pdf.
6. See Geneva Convention relative to the Treatment of Prisoners of War (Third Geneva Convention), Articles 26–32 and 72–75; Geneva Convention rela-

tive to the Protection of Civilian Persons in Time of War (Fourth Geneva Convention), Articles 23, 55–63, and 108–111).

7. See Article 3 common to the four Geneva Conventions; Protocol I, Art. 70.1. Activities of the ICRC are also specified in Articles 9/9/9/10 of the Conventions.

8. http://www.ifrc.org/en/publications-and-reports/code-of-conduct/#sthash.KxloY6Im.dpuf.

9. http://www.sphereproject.org/about/.

10. http://www.ifrc.org/docs/idrl/I522EN.pdf.

11. http://www.peopleinaid.org/pool/files/code/code-en.pdf.

12. http://www.goodhumanitariandonorship.org/gns/principles-good-practice-ghd/overview.aspx.

13. http://www.ifrc.org/en/what-we-do/idrl/idrl-guidelines/.

14. http://www.unisdr.org/.

15. http://www.unisdr.org/2005/wcdr/intergover/official-doc/L-docs/Hyogo-framework-for-action-english.pdf.

16. http://www.recoveryplatform.org/.

17. http://www.gfdrr.org/sites/gfdrr.org/files/publication/GFDRR_Partnership_Charter_2013.pdf.

18. http://tiems.info/About-TIEMS/tiems-code-of-conduct.html.

19. http://ec.europa.eu/echo/policies/disaster_response/mechanism_en.htm.

Chapter 2

1. U.N. Office for the Coordination of Humanitarian Affairs, Philippines: Typhoon-Haiyan, Situation Report No. 17 (as of 25 November 2013).

2. The existence of moral dilemma is debated among philosophers.

3. Thomas Nagel and Isaiah Berlin discuss incommensurable values that lead to moral dilemma. See Nagel (1979) and Berlin (1969).

Chapter 3

1. It is claimed that in Rawls's later work (1993) he re-examined Walzer's perception of distributed goods derived from the "basic intuitive ideas," which are "embedded in the political institutions" of a democratic society (i.e., distinctly political values) (Mulhall and Swift 1996, 207). Thus, Rawls's reconsideration of primary goods as internal to shared social meanings marked the emergence of a methodological convergence between Rawls and Walzer's theories of distributive justice, while the question of the legitimate ground of shared social meanings remains open to debate. For Walzer, reason is always internal to shared social meanings, while for Rawls, shared social meanings are internal to reason.

2. Pogge argues forcefully that the capabilities approach cannot meet the publicity condition of theories of distributive justice, and, in addition, that by identifying some capabilities as more valuable than others it leads to stigmatizing those in society with the less valued capabilities.

3. The term of "personal heterogeneities" was coined by Sen; however, both Sen and Pogge discuss other causes for variation in the ability to convert resources (e.g., climate, environment, gender roles) that are linked to "personal heterogeneities" (Sen 1999a, 70).

4. Pogge's footnote 76 (2002a, 193) reads: "Sen: *Development as Freedom*, 74. This formulation is defective by suggesting that the capability approach features criteria of social justice that take account of the specific ends that different persons are pursuing. This is not the case. Capabilities are defined without regard to such ends. One person does not count as having lesser capabilities than another merely because the former chooses to pursue more ambitious ends. What matters for capability theorists is each person's ability to promote *typical* or *standard human ends*—and not: each person's ability to promote his or her own particular ends."

5. This argument does not imply that Singer does not render the value of personal relationships as a fundamental aspect of human life (2003, 244–45).

Chapter 4

1. For evidence of positive effects of dependency in global public health see Levine (2004).

2. Montreux Retreat X on the Consolidated Appeal Process and Humanitarian Financial Mechanisms. Basic Info, Last updated 12/16/2010.

3. Bill Clinton in testimony before the Senate Foreign Relations Committee, at http://www.democracynow.org/2010/4/1/clinton_rice.

Chapter 5

1. http://www.mofa.go.jp/region/africa/ticad/index.html.
2. http://www.mofa.go.jp/region/africa/ticad/ticad4/index.html.
3. http://unscr.com/en/resolutions/1087.
4. G8, "Gleneagles G8 communique." July 8, 2005. Accessed May 16, 2009. www.g8.gov.uk.
5. https://docs.unocha.org/sites/dms/Documents/TEC_Funding_Report.pdf.
6. http://reliefweb.int/sites/reliefweb.int/files/resources/geo.pdf.
7. Estimated amount of funding according to the UN Office of the Special Envoy for Haiti was $13bn in pledges; almost 50 percent of these pledges ($6bn) have been disbursed. Private donations were estimated at $3bn. See http://www.haitispecialenvoy.org/download/Report_Center/osereport2012.pdf.

8. Michael Robinson Chavez. "Chile Relief Groups Concerned with Donor Fatigue Soon After Haiti Earthquake." *The Star-Ledger*, March 2, 2010. http://www.nj.com/news/index.ssf/2010/03/chile_earthquake_relief_groups.html.

9. For discussion on "emerging donors" in development aid, see Woods, 2008; Reisen, 2007; Villanger, 2007; Manning, 2006.

10. For example, the World Bank's safeguards are set out at http://go.worldbank.org/WTA1ODE7T0. Accessed June 17, 2014.

11. It should be noted that in chapter 3 of *On Economy* in *One World*, Singer is aware of the role of structures of the global economic system in influencing individuals' economic well-being, and the need to change these structures.

Chapter 6

1. U.S. Government Accountability Office, Agency Management of Contractors Responding to Hurricanes Katrina and Rita (February 2006) (GAO-06-461R).

2. United States House of Representatives Committee on Government Reform—Minority Staff Special Investigations Division August 2006, Waste, Fraud, and Abuse in Hurricane Katrina Contracts, pp. 1–19.

3. President's Council on Integrity and Efficiency, Executive Council on Integrity and Efficiency, 10th PCIE Hurricane Katrina Report: To Date as of June 30, 2006; House Government Reform Committee, Hearings on Sifting Through Katrina's Legal Debris: Contracting in the Eye of the Storm (May 3, 2006).

4. GAO, Testimony Before the Committee on Homeland Security and Governmental Affairs, U.S. Senate Hurricanes Katrina and Rita Disaster Relief: Continued Findings of Fraud, Waste, and Abuse, December 6, 2006. http://www.gao.gov/assets/120/115115.pdf.

5. GAO, Expedited Assistance for Victims of Hurricanes Katrina and Rita: FEMA's Control Weaknesses Exposed the Government to Significant Fraud and Abuse, GAO-06-403T. (Washington, DC: GAO, February 13, 2006).

6. http://reliefweb.int/report/haiti/analysis-shows-560-percent-disbursement-rate-haiti-recovery-among-public-sector-donors.

7. http://www.euractiv.com/energy/g8-summit-urges-stringent-nuclea-news-505185.

Chapter 7

1. http://www.euractiv.com/climate-environment/un-climate-talks-start-warnings-news-531615.

2. EM-DAT The OFDA/CRED International Disaster Database. www.em-dat.net—Université Catholique de Louvain—Brussels—Belgium. *Statistical Yearbook for Asia and the Pacific 2011.*

3. Ministry of Environment and Forests Government of the People's Republic of Bangladesh, September 2008. Bangladesh Climate Change Strategy and Action Plan 2008. http://www.moef.gov.bd/moef.pdf.

4. http://unfccc.int/essential_background/convention/background/items/1353.php.

5. http://unfccc.int/essential_background/kyoto_protocol/items/1678.php.

6. For discussion on climate change in cosmopolitan ethics see Caney 2005, Caney 2006, Dobson 2005, Gosseries 2004, 2005, 2007, Elliott 2006, and Jamieson 1992, 1996, 2001.

7. The distinction between adaptation and mitigation costs is highlighted in Caney 2005, 2010. Smit and Pilifosova (2001) by contrast, focus more on "mitigation": see Metz et al., eds. 2001; Henry Shue's illuminating analysis of the different ethical questions raised by global climate change (1993, 1994, 1995).

8. https://unfccc.int/parties_and_observers/parties/annex_i/items/2774.php.

9. http://unfccc.int/resource/docs/convkp/kpeng.html.

10. For that reason, the issue of sharing the burdens of adaption costs has received limited attention in comparison to mitigation. See for example, Paavola and Adger 2006, Grasso 2007, Klinsky and Dowlatabadi 2009, and den Elzen, Meinshausen, and van Vuuren 2007.

11. The intergenerational challenge to the distribution of burdens of global climate change is discussed in Broome 1992, 34–35, and Page 1999.

12. The guidelines of the Brazilian proposal are discussed extensively in Andronova and Schlesinger 2004, den Elzen, Meinshausen, and van Vuuren 2007, Höhne and Blok 2005, and Trudinger and Enting 2005.

13. http://eur-lex.europa.eu/legal-content/EN/ALL/;ELX_SESSIONID=nC79TdXQpQjqcGJ1r872vpBd6lZ2JZYYvYb255ZBQpvKyqrlG2vv!429810554?uri=CELEX:32004L0035.

14. Simon Caney expresses this concern regarding the EPCS principle by suggesting that: "It may be true that some people in the past will have had greater opportunities than some currently living people, but that simply cannot be altered: making their descendants have fewer opportunities will not change that. In fact making their descendants pay for the emissions of previous generations will violate equality, because those individuals will have less than their contemporaries in other countries. So if we take an individualist position, it would be wrong to grant some individuals (those in country A) fewer opportunities than others (those in country B) simply because the people who used to live in country A emitted higher levels of GHGs" (Caney 2008, 689–731, 764–65).

Chapter 8

1. http://www.ifrc.org/en/publications-and-reports/code-of-conduct/.
2. http://www.onphilanthropy.com/bestpract/bp2002-08-16.html.

3. http://www.info.gov.za/gazette/bills/2001/b58-01.pdf.
4. http://www.smartcommunities.ncat.org/wingspread2/wingprin.shtml.
5. http://www.gujaratindia.com/Policies/Policy2.pdf.
6. http://www.itdg.org/?id=disasters_livelihood_approach.
7. It should be noted that communication, training, enforceability, leadership, and ethical climate strategies are not separate from the Code's text.
8. http://www.alnap.org.
9. http://www.compasqualite.org/en/compas-training/index-compas-training.php.
10. http://www.oneworldtrust.org/csoproject/images/documents/INTL8a.pdf.
11. http://www.jointstandards.org/jsiconsultation.

References

Ackerman, John M. *Social accountability in the public sector: A conceptual discussion.* Social Development Papers: Participation and Civic Engagement, Paper No. 82. Washington, DC: World Bank, 2005.

Adams, Guy B., and Danny L. Balfour. *Unmasking administrative evil.* Armonk and London: M. E. Sharpe, 2009.

Adams, J. Stacey, and William B. Rosenbaum. "The relationship of worker productivity to cognitive dissonance about wage inequities." *Journal of Applied Psychology* 46 (1962): 161–64.

Adams, J. Stacey. "Inequity in social exchange." In *Advances in experimental social psychology*, edited by Leonardo Berkowitz, 267–99. New York: Academic Press, 1965.

Aldrich, Daniel P., and Kevin D. Crook. "Strong civil society as a double-edged sword." *Political Research Quarterly* 61 (2008): 379–89.

Andrews, Robert I., and Enzo R. Valenzi. "Overpay inequity or self-image as a worker: A critical examination of an experimental induction procedure." *Organizational Behavior and Human Performance* 5 (1970): 266–76.

Andronova, Natalia, and Michael Schlesinger. "Importance of sulfate aerosol in evaluating the relative contributions of regional emissions to the historical global temperature change." *Mitigation and Adaptation Strategies for Global Change* 9.4 (2004): 383–90.

Archibugi, Daniele. *The global commonwealth of citizens: Toward cosmopolitan democracy.* Princeton, NJ: Princeton University Press, 2008.

Aristotle. *Nichomachean ethics.* Translated by Harris H. Rackham. Cambridge, MA: Harvard University Press, Loeb Classical Library, 1934.

Bangladesh Disaster & Emergency Response (DER) Sub-Group. *Monsoon Floods 2004 Post-Floods Needs Assessment. Summary Report, 30 September 2004.* Dhaka, Bangladesh: DER. http://reliefweb.int/sites/reliefweb.int/files/resources/0601496727BB568AC1256F230033FBC5-lcg-bang-6oct.pdf.

Barcan, Marcus Ruth. "Moral dilemmas and consistency." *Journal of Philosophy* 70 (1980): 121–36.

Barnett, Tim. "Dimensions of moral intensity and ethical decision-making: an empirical study." *Journal of Applied Social Psychology* 31.5 (2001): 1038–57.

Barrett, Lisa F., and Peter Salovey, eds. *The wisdom in feeling*. New York: Guildford, 2002.
Barry, B. "International society from a cosmopolitan perspective." In *International society: Diverse ethical perspectives*, edited by David R. Mapel and Terry Nardin. Princeton, NJ: Princeton University Press, 1998.
Beitz, Charles. "Does global inequality matter?" *Metaphilosophy* 32.1–2 (2001): 95–112.
Beitz, Charles. *Political theory and international relations*, 2nd ed. Princeton, NJ: Princeton University Press, 1979.
Berger, Joseph, M. Hamit Fisek, Robert Z. Norman, and David G. Wagner. "The Formation of Reward Expectations in Status Situations." In *Equity theory: Psychological and sociological perspectives*, edited by David M. Messick and Karen S. Cook, 127–68. New York: Praeger, 1983.
Berlin, Isaiah. "Two concepts of liberty." In *Four essays on liberty*. New York: Oxford University Press, 1969.
Bierhoff, Hans Werner, Ronald L. Cohen, and Jerald Greenberg, eds. *Justice in social relations*. New York: Plenum, 1986.
Blau, Peter M. *Exchange and power in social life*. New York: Wiley, 1964.
Bliss, Christopher, and Rafael Di Tella. "Does competition kill corruption?" *Journal of Political Economy* 105 (1997):1001–23.
Blumstein, Philip W., and Eugene A. Weinstein. "The redress of distributive injustice." *American Journal of Sociology* 74 (1969): 408–18.
Boiral, Olivier. "The certification of corporate conduct: Issues and prospects." *International Labour Review* 142.3 (2003): 317–41.
Borg, Ingwer, and Jim C. Lingoes. *Multidimensional Similarity Structure Analysis*. New York: Springer, 1987.
Borg, Ingwer, and Samuel Shye. *Facet theory: Form and content*. Newbury Park, CA: Sage, 1995.
Bourton, J. (1993) "Recent Trends in the International Relief System," *Disasters* 17.3, September: 187–201.
Bosworth, Barry P., and Susan M. Collins. "Capital flows to developing economies: Implications for saving and investment." *Brookings Papers on Economic Activity* 1 (1999): 143–69.
Brassett, James. "Cosmopolitanism vs. terrorism? Discourses of ethical possibility before and after 7/7." *Millennium: Journal of International Studies* 36.2 (2008): 311–37.
Brickman, Philip, Robert Folger, Erica Goode, and Yaacov Schul. "Microjustice and macrojustice." In *The justice motive in social behavior*, edited by Melvin J. Lerner and Sally C. Lerner, 173–202. New York: Plenum, 1981.
Brighouse, Harry, and Adam Swift. "Legitimate parental partiality." *Philosophy and Public Affairs* 37.1 (2009): 43–80.
Broome, John. *Counting the cost of global warming*. Cambridge: White Horse Press, 1992.

Brown, Roger, and Richard Herrnstein. *Psychology.* Boston, MA: Little Brown, 1975.
BRR/UN. "Tsunami recovery indicators: Status report." December 14, 2005. http://www.humanitarianinfo.org/sumatra/reference/indicators/docs/UNIMS%20Recovery%20Indicators%20Summary_20051214.pdf.
Bueno de Mesquita, Bruce. "Foreign aid and policy concessions." *Journal of Conflict Resolution* 51 (2007): 251–84.
Cahill, Kevin M., ed. *History and hope: The international humanitarian reader.* Bronx, NY: Fordham University Press, 2013.
Caldwell, Lynton C. *International environmental policy: From the twentieth to twenty-first century,* 3rd ed. Durham, NC: Duke University Press, 1996.
Calhoun, Craig. "The imperative to reduce suffering." In *Humanitarianism in question: politics, power, ethics,* edited by Michael Barnett and Thomas G. Weiss, 73–97. Ithaca, NY: Cornell University Press, 2008.
Callaghy, Thomas M. "The state and the development of capitalism in Africa: Theoretical, historical and comparative reflections." In *The precarious balance: State and society in Africa,* edited by Donald Rothchild and Naomi Chazan. Boulder, CO: Westview Press, 1988.
Caney, Simon. "Climate change, justice, and the duties of the advantaged." *Critical Review of International Social and Political Philosophy* 13.1 (2010): 203–28.
Caney, Simon. "Cosmopolitan justice, responsibility, and global climate change." *Leiden Journal of International Law* 18 (2005): 747–75.
Caney, Simon. "Cosmopolitan justice, rights and climate change." *Canadian Journal of Law and Jurisprudence* 19.2 (2006): 255–78.
Caney, Simon. *Cosmopolitan justice, responsibility and global climate change, the global justice reader.* Oxford: Blackwell Publishing, 2008.
Caney, Simon. *Justice beyond borders: A global political theory.* Oxford, UK: Oxford University Press, 2005.
Castells, Manuel. *The information age: Economy, society, and culture. II. The power of identity.* New ed. Oxford, UK: Blackwell, 2003.
Chabal, Patrick, and Jean Pierre Daloz. *Africa works.* London, UK: James Currey, 1999.
Chavla, Leah. "Bill Clinton's heavy hand on Haiti's vulnerable agricultural economy: The American rice scandal." *Council on Hemispheric Affairs,* April 14, 2010.
Cherpitel, Didier. "Deadly Forces." *The Guardian,* March 28, 2001.
Chia, Audrey, and Lim S. Mee. "The effects of issue characteristics on the recognition of moral issues." *Journal of Business Ethics* 27.3 (2000): 255–69.
Clinton, Bill. *Giving: How each of us can change the world.* New York: Knopf, 2007.
Cohen, Ronald L., ed. *Justice: Views from the social sciences.* New York: Plenum, 1986.
Collier, Paul. *The bottom billion.* New York: Oxford University Press, 2007.

Comfort, Louise K., and Thomas W. Haase. "Communication, coherence, and collective action: The impact of Hurricane Katrina on communications infrastructure." *Public Works Management and Policy* 10.3 (2006): 328–43.
Comfort, Louise K., Kilkon Ko, and Adam Zagorecki. "Coordination in rapidly evolving disaster response systems: The role of information." *American Behavioral Scientist* 48 (2006): 295.
Conee, Earl. "Why moral dilemmas are impossible." *American Philosophical Quarterly* 26.2 (1989): 133–41.
Cook, Karen S., and Karen A. Hegtvedt. "Distributive justice, equity, and equality." *Annual Review of Sociology* 9 (1983): 217–41.
Coppola Damon P. *Introduction to international disaster management*, 2nd ed. Oxford, UK: Elsevier/Butterworth-Heinemann, 2011.
Cottingham, John. "Partiality: Favouritism and morality." *Philosophical Quarterly* 36.144 (1986): 357–73.
Cox, Raymond W., III, ed. *Ethics and integrity in public administration. Concepts and cases.* New York: Sharpe, 2009.
CRS. "About Catholic Relief Services." 2007. Accessed August 14, 2013. http://web.archive.org/web/20071012211149/http://crs.org/about/.
Cupit, Geoffrey. "When does justice require impartiality?" Political Studies Association, UK 50th Annual Conference, London, April 10–13, 2000.
Daniell, James E. Damaging Earthquakes Database 2010—The Year in Review, CATDAT, SOS Earthquakes. 2011. http://earthquake-report.com/wp-content/uploads/2011/03/CATDAT-EQ-Data-1st-Annual-Review-2010-James-Daniell-03-03-2011.pdf.
de Waal, Alex. "The humanitarians' tragedy: Escapable and inescapable cruelties." *Disasters* 34.Suppl 2 (2010): S130–7.
Dean, Hartley, ed. *The ethics of welfare: Human rights, dependency and responsibility.* Bristol: The Polity Press, 2004.
Deephouse, David L., and Suzanna Carter. "An examination of differences between organizational legitimacy and organizational reputation." *The Journal of Management Studies* 42.2 (2005): 329–41.
den Elzen, Michel G. J., Jan Fuglestvedt, Niklas Höhne, Cathy Trudinger, Jason Lowe, Ben Matthews, Bård Romstad, Christiano Pires de Campos, and Natalia Andronova. "Analysing countries' contribution to climate change: Scientific and policy-related choices." *Environmental Science & Policy* 8.6 (2005): 614–36.
den Elzen, Michel, Malte Meinshausen, and Detlef van Vuuren. "Multi-gas emission envelopes to meet greenhouse gas concentration targets: Costs versus certainty of limiting temperature increase." *Global Environmental Change* 17.2 (2007): 260–80.
Deutsch, Morton. *Distributive justice: A social-psychological perspective.* New Haven, CT: Yale University Press, 1985.
DFID (Department for International Development). *DFID Annual Report 2007.* London: DFID, 2007.

Dobel, Patrick J. "The corruption of a state." *American Political Science Review* 72.3 (1978): 958–73.
Dobson, Andrew. "Globalisation, cosmopolitanism and the environment." *International Relations* 19.3 (2005): 259–73.
Drabek, Thomas E. *Strategies for coordinating disaster responses*. Boulder, CO: Institute of Behavior Sciences, 2003.
Dreher, Axel, Christos Kotsogiannis, and Steve McCorriston. "Corruption around the world: Evidence from a structural model." *Journal of Comparative Economics* 35.3 (2007): 443–66.
Drèze, Jean, and Amartya Sen. *India: Development and participation*. Oxford, UK: Oxford University Press, 2002.
Eade, Deborah, and Tony Vaux, ed. *Development and humanitarianism: Practical issues*. Bloomfield, NJ: Kumarian Press, Inc., 2007.
Elliott, Lorraine. "Cosmopolitan environmental harm conventions in Global Society." *Cosmopolitanism and Global Institutions* 20.3 (2006): 345–63.
Ellis, Frank. *Rural livelihoods and diversity in developing countries*. Oxford, UK: Oxford University Press, 2000.
Epictetus. *The discourses of Epictetus*. New York: Everyman Paperbacks, 1995.
Erixon, Fredrik, and Sally Razeen. "Policy forum: Economic development: Trade or aid? trade and aid: Countering new millennium collectivism." *The Australian Economic Review* 39.1 (2006): 69–77.
Escaleras, Monica, Nejat Anbarci, and Charles Register. "Public sector corruption and major earthquakes: A potentially deadly interaction." *Public Choice* 132.1 (2007): 209–30.
Fadlalla, Amal Hassan. "The Neoliberalization of Compassion." In *New landscapes of inequality: neoliberalism and the erosion of democracy in America*. Santa Fe, NM: School for Advanced Studies, 2008.
Farmer, Paul. *Haiti after the earthquake*. New York: Public Affairs, 2011.
Fassin, Didier. "Heart of humaneness: The moral economy of humanitarian intervention." In Fassin, D. and Pandolfi, M. (eds). *Contemporary states of emergency: The politics of military and humanitarian interventions*. New York: Zone Books. 2010, pp. 269–95.
Fedderke, Johannes, and Robert Klitgaard. "Economic growth and social indicators: An exploratory analysis." *Journal of Comparative Policy Analysis* 8.3 (2006): 283–303.
Festinger, Leon. *A theory of cognitive dissonance*. Evanston, IL: Row, Peterson, 1957.
Fineman, Martha. "Dependencies." In *Women and welfare: Theory and practice in the United States and Europe*, edited by Nancy Hirschmann and Ulrike Liebert. New Brunswick, NJ: Rutgers University Press, 2001.
Fishbein, Martin, and Icek Ajzen. *Belief, Attitude, intention and behavior: An introduction to theory and research*. Reading, MA: Addison-Wesley, 1975.
Fleischacker, Samuel. *A short history of distributive justice*. Cambridge, MA: Harvard University Press, 2004.

Foa, Uriel G. "Interpersonal and economic resources." *Science* 171 (1971): 345–51.
Foot, Philippa. "Moral realism and moral dilemma." *Journal of Philosophy* 80 (1975): 379–98.
Forgette, Richard, Marvin King, and Bryan Dettrey. "Race, Hurricane Katrina, and government satisfaction: Examining the role of race in assessing blame." *Publius: The Journal of Federalism* 38.4 (2008): 671–91.
Fox, Fiona. "New humanitarianism: Does it provide a moral banner for the 21st century?" *Disasters* 25.4 (2001): 275–89.
Fraser, Nancy. "Reframing justice in a globalizing world." *New Left Review* 36 (2005): 69–88.
Frederickson, H. George. *Ethics and public administration.* New York: M. E. Sharpe, 1993.
Friedrich, James, Paul Barnes, Kathryn Chapin, Ian Dawson, Valerie Garst, and David Kerr. "Psychophysical numbing: When lives are valued less as the lives at risk increase." *Journal of Consumer Psychology* 8 (1999): 277–99.
Gadsden, Amy. "Earthquake rocks civil society." *Far Eastern Economic Review*, June 2008.
Galston, William A. *Justice and the human good.* Chicago, IL: University of Chicago Press, 1980.
GAO. Testimony Before the Committee on Homeland Security and Governmental Affairs, U.S. Senate Hurricanes Katrina and Rita Disaster Relief: Continued Findings of Fraud, Waste, and Abuse. December 6, 2006. http://www.gao.gov/assets/120/115115.pdf.
Gasper, Des. "Ethics and the conduct of international development aid: Charity and obligation." *Forum for Development Studies* 1999/1 (1999): 23–57. Reprinted in Mervyn Frost, ed. *International Ethics.* London, Delhi & Los Angeles: Sage Publications, 2011.
Ghani, Ashraf, Clare Lockhart, and Michael Carnahan. *Closing the sovereignty gap: An approach to state-building.* London: Overseas Development Institute, 2005.
Gibbs, John C. "Kohlberg's moral stage theory: A Piagetian revision." *Human Development* 22 (1979): 89–112.
Giddens, Anthony. *Central problems in social theory.* London: Macmillan Press, 1979.
Gilligan, Carol. *In a different voice: Psychological theory and women's development.* Cambridge, MA: Harvard University Press, 1982.
Gosseries, Axel. "Cosmopolitan luck egalitarianism and the greenhouse effect." *Canadian Journal of Philosophy* 31 (2007): 279–309.
Gosseries, Axel. "Historical emissions and free-riding." In *Justice in time: Responding to historical injustice*, edited by Lukas Meyer, 355–82. Baden-Baden: Nomos, 2004.
Gosseries, Axel. "The egalitarian case against Brundtland's sustainability." *Gaia* 14.1 (2005): 40–46.

Gowans, Christopher. *Innocence lost: An examination of inescapable wrongdoing.* New York: Oxford University Press, 1994.
Grasso, Marco. "A normative ethical framework in climate change." *Climatic Change* 81.3 (2007): 223–46.
Greenberg, Jerald, and Ronald L. Cohen. "Why justice? Normative and instrumental interpretations." In *Equity and justice in social behavior*, edited by Jerald Greenberg and Ronald L. Cohen, 437–69. New York: Academic Press, 1982.
Guha-Sapir, Debby, F. Vos, and R. Below. *Annual disaster statistical review 2011: The numbers and trends.* Brussels: Center for Research on the Epidemiology of Disasters, 2012.
Habermas, Jürgen. "Moral development and ego identity." In *Communication and the evolution of society*, edited by J. Habermas, 69–94. Boston, MA: Beacon Press, 1979.
Hall, Peter. "Public policy-making as social resource creation." *APSA-CP* 16.2 (2005): 1–4.
Hammer, Joshua. "Inside the danger zone." *Newsweek*, April 11, 2011, 28–31.
Harding, Carol Gibb, ed. *Moral dilemmas: Philosophical and psychological issues in the development of moral reasoning.* Chicago, IL: Precedent Publishing Inc., 1985.
Harris, Paul G. *World ethics and climate change: from international to global justice.* Edinburgh: Edinburgh University Press, 2010.
Harvey, Paul, and Jeremy Lind. "Dependency and humanitarian relief: A critical analysis." Overseas Development Institute. HPG Research Report. July 19, 2005.
Hawley, Chris. "Haiti drug trafficking likely to rise in quake aftermath." *USA Today*, February 9, 2010. http://www.usatoday.com/news/world/2010-02-09-haiti-drugtrafficking_N.htm.
Hegtvedt, Karen A., and Barry Markovsky. "Justice and injustice." In *Sociological perspectives on social psychology*, edited by Karen S. Cook, Alan Gary Fine, and James S. House, 257–280. Boston, MA: Allyn and Bacon, 1995.
Hegtvedt, Karen A., and Cathy Johnson. "A justice beyond the individual: A future with legitimation." *Social Psychology Quarterly* 63 (2000): 298–311.
Heidenheimer, Arnold J. "Perspectives on the perception of corruption." In *Political corruption*, edited by Arnold J. Heidenheimer and Michael Johnston, 144–54. New Brunswick, NJ: Transaction Publishers, 2002.
Heidenheimer, Arnold J., Michael Johnston, and Victor T. Levine, eds. *Political corruption.* New Brunswick: Transaction Publishers, 1989.
Heider, Fritz. "Attitudes and cognitive organization." *Journal of Psychology* 21 (1946): 107–12.
Heider, Fritz. *The psychology of interpersonal relations.* New York: Wiley, 1958.
Heilbroner, Robert L., ed. "The Theory of Moral Sentiments." In *The essential Adam Smith.* New York: W. W. Norton & Company, 1987.

Held, David. "Cosmopolitanism; globalization tamed?" *Review of International Studies* 29 (2003): 465–80.
Held, David. *Democracy and the global order: From the modern state to cosmopolitan governance.* Cambridge, UK: Polity Press, 1995.
Henson, Spencer, Johanna Lindstrom, and Lawrence Haddad. "Public perceptions of international development and support for aid in the UK: Results of a qualitative enquiry." Institute of Development Studies, 2010. http://www.ids.ac.uk/download.cfm?objectid=9EBBB455-0CF8-6390-DAD44D86E2467C3
Höhne, Niklas, and Kornelis Blok. "Calculating historical contributions to climate change—discussing the 'Brazilian Proposal.'" *Climatic Change* 71.1 (2005): 141–73.
Holmberg, Sören, Bo Rothstein, and Naghmeh Nasiritousi. "Quality of government: What you get." *Annual Review of Political Science* 12 (2009): 135–61.
Homans, George C. "Social behavior as exchange." *American Journal of Sociology* 63 (1958): 447–58.
Homans, George C. *Social behavior: Its elementary forms.* New York: Harcourt, Brace, 1961.
Homans, George C. *Social behavior: Its elementary forms.* New York: Harcourt, Brace, 1974.
Howell, Colonel Willis. Interview 18 November 2013. "Charities worry about 'donor fatigue' after recent disasters." http://www.wsoctv.com/news/news/local/charities-worry-about-donor-fatigue-after-recent-d/nbw2B/.
Hulme, David. *Global poverty: How global governance is failing the poor.* London: Routledge, 2010.
Hume, David. *A treatise of human nature.* 2nd ed., edited by Sir Lewis Amherst Selby-Bigge, with an analytical index. Oxford, UK: Clarendon, 1978.
Ink, Dwight. "An analysis of the House Select Committee and White House Reports on Hurricane Katrina." *Public Administration Review* 66.6 (2006): 800–7.
Intergovernmental Panel on Climate Change (IPCC). *Climate change 2001*, Third Assessment Report, *Climate Change 2001: Mitigation.* London: Cambridge University Press, 2001.
Intergovernmental Panel on Climate Change (IPCC). *Climate change 2001: Synthesis report.* Cambridge, MA: Cambridge University Press, 2002.
Intergovernmental Panel on Climate Change (IPCC). *Climate change 2007: Synthesis report.* Cambridge, MA: Cambridge University Press, 2007a.
Intergovernmental Panel on Climate Change (IPCC). *IPCC Fourth Assessment Report: Summary for policymakers of the synthesis report.* 2007b. http://www1.ipcc.ch/pdf/assessment-report/ar4/syr/ar4_syr_spm.pdf.
International Committee of the Red Cross (ICRC). Statutes of the International Red Cross and the Red Crescent Movement, 1986, as amended in 1995 and 2006. Accessed May 16, 2012. http://www.icrc.org/eng/assets/files/other/statutes-en-a5.pdf&.

Ireni Saban, Liza. "Understanding the Obligations of Codes of Ethics." In *Handbook of public administration*, 3rd ed, edited by James Perry and Rob Christensen. Hoboken, NJ: Jossey-Bass, 2014.

Ireni Saban, Liza. *Disaster emergency management: The emergence of professional help for victims of natural disasters.* New York: SUNY Press, 2014.

Jamieson, Dale. "Climate change and global environmental justice." In *Changing the atmosphere: Expert knowledge and environmental governance*, edited by Clark A. Miller and Paul N. Edwards. Cambridge, MA: MIT Press, 2001.

Jamieson, Dale. "Ethics and intentional climate change." *Climatic Change* 33 (1996): 323–36.

Jamieson, Dale. "Ethics, public policy and global warming." *Science, Technology and Human Values* 17 (1992): 139–53. Reprinted in Dale Jamieson, *Morality's progress*. Oxford: Oxford University Press, 2003.

Jenni, Karen E., and George Loewenstein. "Explaining the 'identifiable victim effect.'" *Journal of Risk and Uncertainty* 14 (1997): 235–57.

Jeong, Changwoo, and Hyemin Han. "Exploring the relationship between virtue ethics and moral identity." *Ethics & Behavior* 23.1 (2013): 44–56.

Jia, Xijin. "Chinese civil society after the May 12 earthquake." English translation on Policy Forum Online 08-056A. (July 22, 2008). www.nautilus.org/fora/security/08056Jia.html.

Johnson, Phil, and Ken Smith. "Contextualizing business ethics: Anomie and social life." *Human Relations* 52.11 (1999): 1351–76.

Jones, Joanne, Dawn W. Massey, and Linda Thorne. "Auditors' ethical reasoning: Insights from past research and implications for the future." *Journal of Accounting Literature* 22 (2003): 45–103.

Kapucu, Naim, Marie Eelena Augustin, and Vener Garayev. "Interstate partnerships in emergency management: Emergency Management Assistance Compact (EMAC) in response to catastrophic disasters." *Public Administration Review* 69.2 (2009): 297–313.

Kapucu, Naim, Tolga Arslan, and Lloyd M. Collins. "Examining intergovernmental and interorganizational response to catastrophic disasters: Toward a network-centered approach." *Administration & Society* 42 (2010): 222–47.

Kapucu, Naim. "Interagency communication networks during emergencies: Boundary spanners in multiagency coordination." *American Review of Public Administration* 36 (2006): 207–25.

Kapucu, Naim. "Lessons of disaster: Policy change after catastrophic events." *Public Management Review* 10.1 (2008): 153–55.

Kasher, Asa. "Professional ethics." In *Ethical issues in mental health consultation and therapy*, edited by Gabi Scheffler, Judith Achmon, and Gabriel Weil, 15–29. Jerusalem: Magnes, 2003.

Kaufmann, Daniel, Aart Kraay, and Massimo Mastruzzi. *Governance matters 2009: Learning from over a decade of the worldwide governance indica-*

tors. Washington, DC: Brookings, 2009. http://www.brookings.edu/opinions/2009/0629_governance_indicators_kaufmann.aspx

Kayser, Egon, and Thomas Schwinger. "A theoretical analysis of the relationship among individual justice concepts, layman psychology and distribution decisions." *Journal for the Theory of Social Behavior* 12 (1982): 47–51.

Kellett, Jan, and Dan Sparks. "Disaster risk reduction: Spending where it should count. Global humanitarian assistance." *Development Initiatives: Bristol*, 2012: 1–40.

Kenny, Charles. "Haiti doesn't need your old t-shirt." *Foreign Policy*, November 10, 2011.

Kimberly, James C. "The emergence and stabilization of stratification in simple and complex social systems." *Sociological Inquiry* 40 (1970): 73–101.

Klinsky, Sonja, and Hadi Dowlatabadi. "Conceptualizations of justice in climate policy." *Climate Policy* 9 (2009): 88–108.

Klitgaard, Robert, Johannes Fedderke, and Kamil Akramov. "Choosing and using performance criteria." In *High-performance government: structure, leadership, incentives*, edited by Robert Klitgaard and Paul C. Light, 407–46. Santa Monica, CA: The RAND Corporation, 2005. http://www.rand.org/pubs/monographs/2005/RAND_MG256.pdf.

Klitgaard, Robert. "Subverting corruption." *Global Crime* 7.3-4 (2006): 299–307. http://www.cgu.edu/PDFFiles/Presidents%20Office/Subverting_Corruption_8-06.pdf.

Kogut, Tehila, and Ilana Ritov. "The singularity of identified victims in separate and joint evaluations." *Organizational Behavior and Human Decision Processes* 97 (2005b): 106–16.

Kogut, Tehile, and Ilana Ritov. "The 'identified victim' effect: An identified group, or just a single individual?" *Journal of Behavioral Decision Making* 18 (2005a): 157–67.

Kohlberg, Lawrence. "Stage and sequence: the cognitive-developmental approach to socialization." In *Handbook of socialisation theory and research*, edited by: David A. Goslin, 347–480. Chicago, IL: Rand McNally, 1969.

Kohlberg, Lawrence. *The philosophy of moral development*. San Francisco, CA: Harper and Row, 1981.

Koliba, Christopher, Jack Meek, and Asim Zia. *Governance networks in public administration and public policy*. Boca Raton, FL: CRC Press, 2010.

Krasner, Stephen. *International regimes*. Ithaca, NY: Cornell University Press, 1983.

Kurer, Oskar. "Corruption: An alternative approach to its definition and measurement." *Political Studies* 53.1 (2005): 222–39.

Laframboise, Nicole, and Boileau Loko. "Natural disasters: Mitigating Impact, managing risks." Working Paper, WP/12/245. Washington, DC: International Monetary Fund, 2012. http://www.imf.org/external/pubs/ft/wp/2012/wp12245.pdf.

Lancaster, Carol. "Aid effectiveness in Africa: The unfinished agenda." *Journal of African Economies* 8.4 (1999): 487–503.

Lane, Irving M. and Lawrence A. Mess. "Distribution of insufficient, sufficient and oversufficient rewards: a clarification of equity theory." *Journal of Personality and Social Psychology* 21 (1972): 288–333.

Lane, Irving M., and Lawrence A. Mess. "Equity and the distribution of rewards." *Journal of Personality and Social Psychology* 20.1 (1971): 1–17.

Lappe, Frances M., and Joseph Collins. *Food first*. New York: Ballantine Books, 1977.

Lasswell. Harold D. Polities: Who gets what, when, how. New York: McGraw-Hill, 1936.

Lawler, Edward E. III. "Equity theory: A predictor of productivity and work quality." *Psychological Bulletin* 70 (1968): 596–610.

Lensink, Robert, and Howard White. *Aid dependence. Issues and indicators*. EGDI Study 1999:2, Stockholm: Ministry for Foreign Affairs, 1999.

Lenski, Gerhard. *Power and privilege*. San Francisco, CA: McGraw-Hill, 1966.

Lentz, Erin, and Christopher B. Barrett. *Food aid targeting shocks and private transfers among East African pastoralists*. Working Paper. Ithaca, NY: Cornell University, 2005. http://www.cfnpp.cornell.edu/images/wp170.pdf.

Lerner, Melvin J., and Linda A. Whitehead. "Procedural justice viewed in the context of justice motive theory." In *Justice in social interaction*, edited by Gerold Mikula, 219–56. New York: Springer Verlag, 1980.

Leventhal, Gerald S. "The distribution of rewards and resources in groups and organizations." In *Advances in experimental social psychology*, edited by Leonard Berkowitz and Elaine Walster, vol. 9: 91–131. New York: Academic Press, 1976.

Leventhal, Gerald S. "What should be done with equity theory? New approaches to the study of fairness in social relationships." In *Social exchange. Advances in theory and research*, edited by Kenneth J. Gergen, Martin S. Greenberg, and Richard H. Willis, 27–55. New York: Plenum Press, 1980.

Leventhal, Gerald S., and James W. Michaels. "Extending the equity model: Perceptions of inputs and allocations of reward as a function of duration and quantity of performance." *Journal of Personality and Social Psychology* 12 (1969): 303–9.

Leventhal, Gerald S., Jurgis Karuza, and William Rick Fry. "Beyond fairness: A theory of allocation preferences." In *Justice and social interaction*, edited by Gerold Mikula, 167–218. New York: Plenum Press, 1980.

Levine, Ruth. *Millions Saved: Proven successes in global health*. Washington DC: Center for Global Development, 2004.

Lewin, Kurt. *A dynamic theory of personality*. New York: McGraw-Hill, 1935.

Lewin, Kurt. *Principles of topological psychology*. New York: McGraw-Hill, 1936.

Lifton, Robert J. *Death in life: Survivors of Hiroshima*. New York: Random House, 1967.

Lind, E. Allen, and Tom R. Tyler. *The social psychology of procedural justice*. New York: Plenum Press, 1988.

Long, Brad S., and Cathy Driscoll. "Codes of ethics and the pursuit of organizational legitimacy: theoretical and empirical contributions." *Journal of Business Ethics* 77 (2008): 173–89.

Manning, Richard. "Will 'emerging donors' change the face of international cooperation?" *Development Policy Review* 24.4 (2006): 371–85.

McEntire, David A. "Coordinating multi-organizational responses to disaster: Lessons from the March 28, 2000, Fort Worth Tornado." *Disaster Prevention and Management* 11.5 (2002): 369–79.

Meeker, Barbara F. "Decisions and exchange." *American Sociological Review* 36 (1971): 485–95.

Menzel, Donald. "The Katrina aftermath: A failure of federalism or leadership?" *Public Administration Review* 66.6 (2006): 808–12.

Metz, Bert, Ogunlade Davidson, Rob Swart, and Jahua Pan, eds. *Climate change 2001: Mitigation—contribution of Working Group III to the third assessment report of the intergovernmental Panelon climate change*. 2001.

Migdal, Joel. *State in society: Studying how states and societies transform and constitute each other*. New York: Cambridge University Press, 2001.

Mikula, Gerold, and Thomas Schwinger. "Intermember relations and reward allocation: Theoretical considerations of affects." In *Dynamics of group decisions*, edited by Hermann Brandstätter, James H. Davis, and Heinz Schuler, 229–50. Beverly Hills, CA: Sage, 1978.

Mikula, Gerold. "Beyond fairness: A theory of allocation preferences." In *Justice and social interaction*, edited by Gerold Mikula, 167–218. New York: Plenum Press, 1980.

Miller, David. "Distributive justice: What the people think." *Ethics* 102 (1992): 555–59.

Mitchell, James K. "The primacy of partnership: Scoping a new national disaster recovery policy." *Annals of the American Academy of Political and Social Science* 604 (2006): 228–55.

Mohamad, Abu Kassim Bin. "effective anti-corruption enforcement and complaint-handling mechanisms: The Malaysian experience." In *Curbing corruption in tsunami relief operations*. Proceedings of the Jakarta Expert Meeting, Jakarta, April 7–8, 2005. http://www. u4.no/document/literature/adb-ti-2005-curbing-corruption-tsunami-relief-operations.pdf.

Montenegro, Alvaro, Victor Brovkin, Michael Eby, David Archer, and Andrew J. Weaver. "Long-term fate of anthropogenic carbon." *Geophysical Research Letters* 34 (2007): 1–5.

Mooney, Gerry. "'Problem' populations, 'problem' places." In *Social justice: Welfare, crime and society*, edited by J. Neman and N. Yeats. Maidenhead: Open University Press, 2009.

Mosley, Paul. "The political economy of foreign aid: A model of the market for a public good." *Economic Development and Cultural Change* 33 (1985): 373-94.

Mowday, Richard T. "Equity theory predictions of behavior in organizations." In *Motivation and leadership at work*, edited by Lyman Porter, Gregory Bigley, and Richard Steers, 53-71. New York: McGraw-Hill, 1996.

Moynihan, Donald P. "Extra-network organizational reputation and blame avoidance in networks: The Hurricane Katrina example." *Governance* 25.4 (2012): 567-88.

Mulhall, Stephen, and Adam Swift. *Liberals and Communitarians*, 2nd ed. Oxford, UK: Blackwell, 1996.

Munck, Ronaldo. *Globalization and social exclusion: A transformationalist perspective*. Bloomfield, CT: Kumarian Press, 2005.

Nagel, Thomas. "The fragmentation of value." In *Mortal questions*. Cambridge, UK: Cambridge University Press, 1979.

Naím, Moisés. "Rogue aid." *Foreign Policy*, online, March/April 2007: 95-96.

Najam, Adil, Saleemul Huq, and Youba Sokona. "Climate negotiations beyond Kyoto: Developing countries concerns and interests." *Climate Policy* 3.3 (2003): 221-31.

Neumayer, Eric. "In defence of historical accountability for greenhouse gas emissions." *Ecological Economics* 33 (2000): 185-92.

Nickel, Patricia M., and Angela M. Eikenberry. "Responding to 'natural' disasters: The ethical implications of the voluntary state." *Administrative Theory & Praxis* 29.4 (2007): 534-45.

Nickerson, Colin. "Relief workers shoulder a world of conflict; aid agencies encounter growing dangers as nations withhold peacekeeping troops." *Boston Globe*, July 27, 1997.

Nolte, Isabella M., Eric C. Martin, and Silke Boenigk. "Cross-sectoral coordination of disaster relief." *Public Management Review* 14.6 (2012): 707-30.

Northouse, Peter G. *Leadership: Theory and practice*. Thousand Oaks, CA: Sage Publications, 2001.

Nye, Joseph S. "Corruption and political development: A cost-benefit analysis." *American Political Science Review* 61 (1967): 417-27.

O'Donovan, Gary. "Environmental disclosures in the annual report: Extending the applicability and predictive power of legitimacy theory." *Accounting, Auditing and Accountability Journal* 15.3 (2002): 344-72.

OECD DAC. 2006 *Development co-operation report*. Paris: OECD, 2007.

OECD. *Development co-operation report 2010*. Paris: OECD, 2010.

OECD. *Public opinion and the fight against poverty*. Paris: OECD, 2003.

OECD/PUMA. *Ethics in the public service: Current issues and practice*, Paper No. 14. Paris: OECD, 1996.

Olivier Jos G. J., Janssens-Maenhout Greet, Muntean Marilena and Peters Jeroen A. H. W. *Trends in global CO2 emissions: 2014 report*, Netherlands Environmental Assessment Agency, The Hague. 2014.

Opsahl, Robert L. and Marvin D. Dunnette. "The role of financial compensation in industrial motivation." *Psychological Bulletin* 66 (1966): 94–118.

Orbinski, James. " 'No doctor can stop genocide': Doctors Without Borders' humanitarian actions 'cannot erase the long-term necessity of political responsibility.'" *The Ottawa Citizen*, December 17, 1999.

Özerdem, Alpaslan, and Tim Jacoby. *Disaster management and civil society: Earthquake relief in Japan, Turkey and India*. London, New York: I. B. Tauris, 2006.

Paavola, Jounni, and Neil Adger. "Fair adaptation to climate change." *Ecological Economics* 56.4 (2006): 594–609.

Page, Edward. "Intergenerational justice and climate change." *Political Studies* 1.47 (1999): 61–6.

Pawson, Lara. "Angola calls a halt to IMF talks." *BBC News*, March 13, 2007. Accessed August 26, 2008. http://news.bbc.co.uk/i/hi/busi ness/6446025.stm.

Pogge Thomas. "An egalitarian law of peoples." *Philosophy and Public Affairs* 23.3 (1994): 195–224.

Pogge Thomas. *World poverty and human rights*. Cambridge, UK: Polity Press, 2002b.

Pogge, Thomas W. " 'Assisting' the global poor." In *The ethics of assistance: Morality and the distant needy*, edited by Deen K. Chaterjee, 260–88. Cambridge, UK: Cambridge University Press, 2004.

Pogge, Thomas W. "Can the capability approach be justified?" 2002. Accessed May 16, 2014. http://philosophyfaculty.ucsd.edu/faculty/rarneson/courses/pogge1capability.pdf, 1–71.

Pogge, Thomas W. "Can the capability approach be justified?" In *Global inequalities*, edited by Martha Nussbaum and Chad Flanders, special issue 30.2 (2002a) (appeared February 2004 in *Philosophical Topics*, 167–228). Also available at http://www.mit.edu/~shaslang/mprg/PoggeCCABJ.pdf, 1–70.

Pogge, Thomas W. "Eradicating Systemic Poverty: Brief for a global resources dividend." *Journal of Human Development* 2.1 (2001): 59–77.

Pogge, Thomas. "A global resources dividend." In *ethics of consumption: The good life, justice, and global stewardship*, edited by David A. Crocker and Toby Linden. Lanham, MD: Rowman & Littlefield, 1998.

Pogge, Thomas. "Cosmopolitanism and sovereignty." *Ethics* 103.1 (1992): 48–75.

Pogge, Thomas. "Moral priorities for international human rights NGOs." In *Ethics in action: The ethical challenges of international human rights nongovernmental organizations*, edited by Daniel A. Bell and Jean-Marc Coicaud. Cambridge University Press, New York, 2007.

Pogge, Thomas. *Realizing Rawls*. Ithaca, NY, and London: Cornell University Press, 1989.
Pogge, Thomas. *World poverty and human rights*. Malden, MA: Polity Press, 2008.
Polman, Linda. *The Crisis Caravan: What's Wrong with Humanitarian Aid?* New York: Metropolitan Books, 2010.
President's Council on Integrity and Efficiency, Executive Council on Integrity and Efficiency. "10th PCIE Hurricane Katrina Report: To Date as of 30 June 2006."
Rawls, John. *A theory of justice*. Oxford: Clarendon Press, 1972.
Rawls, John. *Political liberalism*, New York: Columbia University Press, 1993.
Reimann, Kim. "A view from the top: international politics, norms and the worldwide growth of NGOs." *International Studies Quarterly* 50.1 (2006): 45–68.
Reisen, Helmut. *Is China actually helping improve debt sustainability in Africa?* Paris: OECD Development Centre, July 2007.
Rescher, Nicholas. *Distributive justice*. New York: Bobbs-Merrill, 1966.
Rest, James R. *Moral development: Advances in research and theory*. New York: Praeger, 1986.
Reynolds, Scott. "Moral awareness and ethical predispositions: Investigating the role of individual differences in the recognition of moral issues." *Journal of Applied Psychology* 91 (2006): 233–43.
Ridell, Roger, and Rehman Sobhan. *Aid dependency: Causes, symptoms and remedies*. SIDA Project 2015, 1996.
Rieff, David. *A bed for the night: Humanitarianism in crisis*. New York: Simon and Schuster, 2002.
Rodan, Garry, and Caroline Hughes. "Ideological coalitions and the international promotion of social accountability: The Philippines and Cambodia compared." *International Studies Quarterly* 56 (2012): 367–80.
Roemer, John. *Theories of distributive justice*. Ithaca, NY: Cornell University Press, 1993.
Rose, Adam. "Economic principles, issues, and research priorities in hazard loss estimation." In *Economic principles, issues, and research priorities in hazard loss estimation*, edited by Yasuhide Okuyama and Stephanie Chang, 13–36. Heidelberg, Germany: Springer, 2004.
Ross, William David. "What makes right acts right?" In *The right and the good*. Oxford, UK: Clarendon Press, 1930.
Rothstein, Bo, and Jan Teorell. "What is quality of government: A theory of impartial political institutions." *Governance: An International Journal of Policy and Administration* 21.2 (2008): 165–90.
Rubenstein, Jennifer. "Distribution and emergency." *The Journal of Political Philosophy* 15.3 (2007): 296–320.
Rubenstein, Jennifer. "Humanitarian NGOs' duties of justice." *Journal of Social Philosophy* 40.4 (2009): 524–41.

Rubenstein, Jennifer. "The Distributive commitments of international NGOs." In *Humanitarianism in question*, edited by Michael Barnett and Thomas G. Weiss. Ithaca, NY: Cornell University Press, 2008.

Rubenstein, Jennifer. "The misuse of power, not bad representation: Why it is beside the point that no one elected Oxfam." *The Journal of Political Philosophy* 22.2 (2014): 204–30.

Rubenstein, Jennifer. *Between Samaritans and states: The political ethics of humanitarian INGOs*. Oxford: Oxford University Press, 2015.

Rubinstein, David. "The concept of justice in sociology." *Theory and Society* 17 (1988): 527–50.

Sabbagh, Clara, Yechezkel Dar, and Nura Resh. "The structure of social justice judgments: A facet approach." *Social Psychology Quarterly* 57 (1994): 244–61.

Sampson, Edward E. "On justice as equality." *Journal of Social Issues* 31 (1975): 45–64.

Sandbrook, Richard. *The politics of Africa's economic recovery*. Cambridge, UK: Cambridge University Press, 1992.

Sartre, Jean Paul. "Existentialism is a humanism." Translated by Bernard Frechtman. In *Ethics: History, theory, and contemporary issues*, edited by Steven M. Cahn and Peter Markie. Oxford, UK: Oxford University Press, 2002.

Scheffler, Samuel. *Boundaries and allegiances*. Oxford, UK: Oxford University Press, 2001.

Schelling, Thomas C. "The life you save may be your own." In *Problems in public expenditure*, edited by Samuel B. Chase, Jr., 127–76. Washington, DC: Brookings Institute, 1968.

Schuller, Mark. "Looking at Haiti from Haiti: Two years after the earthquake, a new book aims to tell the story we've missed." *Boston Globe*, January 15, 2012.

Scott, Richard. *Institutions and organizations*. Thousand Oaks, CA: Sage Publications, 1995.

Sen, Amartya, "Justice: Means versus freedom." *Philosophy & Public Affairs* 19 (1990): 111–21.

Sen, Amartya, *Inequality reexamined*. Cambridge, MA: Harvard University Press, 1992.

Sen, Amartya. "Human rights and capabilities." *Journal of Human Development* 6 (2005): 151–66.

Sen, Amartya. "Positional objectivity." *Philosophy and Public Affairs* 22 (1993): 126–45.

Sen, Amartya. "The standard of living." In *The standard of living: The Tanner lectures: Cambridge, 1985*, edited by Amartya Sen, John Muellbauer, Ravi Kanbur, Keith Hart, and Bernard Williams. Cambridge, UK: Cambridge University Press, 2000.

Sen, Amartya. "Well-Being, agency and freedom. The Dewey lectures 1984." *The Journal of Philosophy* 82.4 (1985): 169–221.

Sen, Amartya. *Commodities and capabilities*, New Delhi, India: Oxford University Press, [1985] 1999a.
Sen, Amartya. *Commodity and capabilities*. Amsterdam, The Netherlands: Elsevier Science, 1985.
Sen, Amartya. *Development as freedom*. New York: Knopf, 1999.
Sen, Amartya. *Development as freedom*. New York: Oxford University Press, 1999.
Sen, Amartya. *Rationality and freedom*, Cambridge and London: Harvard University Press, 2002.
Sen, Amartya. *Reason before identity. The Romanes lecture for 1998*. New Delhi, Oxford, New York: Oxford University Press, 1999b.
Sen, Amartya. *Resources, values, and development*. Cambridge, MA: Harvard University Press, 1984.
Sen, Amartya. *The idea of justice*. Cambridge, MA: Harvard University Press, 2009.
Sen. Amartya. "The ends and means of sustainability." *Journal of Human Development and Capabilities: A Multi-Disciplinary Journal for People-Centered Development* 14.1 (2013): 6–20.
Shue, Henry. "After you: May action by the rich be contingent upon action by the poor?" *Indiana Journal of Global Legal Studies* 1 (1994): 344.
Shue, Henry. "Avoidable necessity: Global warming, international fairness, and alternative energy." *Theory and Practice: NOMOS XXXVII*, edited by I. Shapiro and J. Wagner de Cew, 240. New York: New York University Press, 1995.
Shue, Henry. "Subsistence emissions and luxury emissions." *Law and Policy* 15 (1993): 40.
Singer, Peter. "A response." In *Singer and his critics*, edited by Dale Jamieson. London: Blackwell, 1999: 269–336.
Singer, Peter. "Famine, affluence and morality." *Philosophy & Public Affairs* 1.3 (1972): 229–43.
Singer, Peter. "Moral experts." (1972b). Reprinted in *Writings on an ethical life*. New York: Ecco Press, 2000: 3–6.
Singer, Peter. "The drowning child and the expanding circle." *New Internationalist* April 1997, Issue 289.
Singer, Peter. "The Singer solution to world poverty." Reprinted in *Writings on an ethical life*. London: Fourth Estate, 1999: 118–24.
Singer, Peter. *One world: The ethics of globalization*. New Haven, CT: Yale University Press, 2004.
Singer, Peter. *One world: The ethics of globalization*. Princeton, NJ: Princeton University Press, 2002.
Singer, Peter. *Practical ethics*. Cambridge, UK: Cambridge University Press, 2003.
Singer, Peter. *The life you can save: Acting now to end world poverty*. New York: Random House, 2009.
Singhapakdi, Anusorn, Scott J. Vitell, and Kenneth L. Kraft. "Moral intensity and ethical decision-making of marketing professionals." *Journal of Business Research* 36.3 (1996): 245–55.

Sinnott-Armstrong, Walter. *Moral dilemmas*. New York: Basil Blackwell, 1988.
Slim, Hugo. "Doing the right thing: Relief agencies, moral dilemmas and moral responsibility in war and political emergencies." *Disasters* 21.3 (1997): 244–57.
Slim, Hugo. "Not philanthropy but rights: The proper politicisation of humanitarian philosophy." *The International Journal of Human Rights* 6.2 (2002): 1–22.
Slim, Hugo." International NGOs—a necessary good. global insight non-governmental organisations," (2013). Available at http://www.hugoslim.com/Pdfs/A%20Necessary%20Good%20.pdf.
Slovic, Paul, David Zionts, Andrew K. Woods, Ryan Goodman, and Derek Jinks. "Psychic numbing and mass atrocity." In *The behavioral foundations of public policy*, edited by Eldar Shafir, 126–42. Princeton, NJ: Princeton University Press, 2013.
Slovic, Paul, Melissa L. Finucane, Ellen Peters, and Donald G. MacGregor. "The affect heuristic." In *Heuristics and biases: The psychology of intuitive judgment*, edited by Thomas Gilovich, Dale Griffin, and Daniel Kahneman, 397–420. New York: Cambridge University Press, 2002.
Small, Debora A., and George Loewenstein. "Helping a victim or helping the victim: Altruism and identifiability." *Journal of Risk and Uncertainty* 26 (2003): 5–16.
Small, Debora A., and George Loewenstein. "The devil you know: The effects of identifiability on punishment." *Journal of Behavioral Decision Making* 18 (2005): 311–18.
Small, Debora A., George Loewenstein, and Paul Slovic. "Sympathy and callousness: The impact of deliberative thought on donations to identifiable and statistical victims." *Organizational Behavior and Human Decision Processes* 102 (2007): 143–53.
Smillie, Ian, and Larry Minear. *Charity of nations: Humanitarian action in a calculating world*. Bloomfield, NJ: Kumarian Press, Inc., 2004.
Smit, Barry, and Olga Pilifosova. "Adaptation to climate change in the context of sustainable development and equity." In *Climate change 2001: Impacts, adaptation and vulnerability*, edited by James J. McCarthy, Intergovernmental Panel on Climate Change. Working Group II, Chapter 18, 877–912. Cambridge, UK: Cambridge University Press, 2001.
Smith, Steven R. "The challenge of strengthening nonprofits and civil society." *Public Administration Review* 68 (2008): S132–45.
Snell, Robin S. "Complementing Kohlberg: mapping the ethical reasoning used by managers for their own dilemma cases." *Human Relations* 49.1 (1996): 23–49.
Spektorowski, Alberto, and Liza Ireni Saban. *Politics of eugenics: Productionism, population, and national welfare*. Oxford, UK: Routledge, 2013.
Srinivasan, U. Thara, Susan P. Carey, Eric Hallstein, Paul A. T. Higgins, Amber C. Kerr, Laura E. Koteen, Adam B. Smith, Reg Watson, John Harte, and

Richard B. Norgaard. "The debt of nations and the distribution of ecological impacts from human activities." *Proceedings of the National Academy of Sciences* 105.5 (2008): 1768–73.

Stiglitz, Joseph. *Globalisation and its discontents*. New York: W. W. Norton, 2002.

Stivers, Camilla. "So poor and so black: Hurricane Katrina, public administration, and the issue of race." *Public Administration Review*, Special Issue December (2007): 48–56.

Stoddard, Abby, and Adele Harmer. "Supporting security for humanitarian action: A review of critical issues for the humanitarian community." Montreux X conference, March 2010. http://www.humanitarianoutcomes.org/sites/default/files/resources/SupportingSecurityforHumanitarianActionMarch20101.pdf.

Stoddard, Abby, Adele Harmer, and Katherine. Haver. *Providing aid in insecure environments: Trends in policy and operations*. London: Overseas Development Institute, 2006.

Streeten, Paul P. "Basic needs: Premises and promises." *Journal of Policy Modeling* 1 (1979): 136–46.

Streeten, Paul P., Shahid J. Burki, Mahbub ul Haq, Norman Hicks, and Frances Stewart. *First things first, meeting basic human needs in developing countries*. Oxford, UK and New York: Oxford University Press, 1981.

Styron, William. *Sophie's choice*. New York: Bantam, 1980.

Suchman, Mark. "Managing legitimacy: Strategic and institutional approaches." *Academy of Management Review* 20 (1995): 571–610.

Sullivan, Edmund V. "A study of Kohlberg's structure theory of moral development: A critique of liberal social science ideology." *Human Development* 20 (1977): 352–76.

Tabuchi, Hiroko, Ken Belson, and Normitsu Onishi. "Warnings ignored at crippled reactor." *International Herald Tribune*, March 23, 2011.

Tan, Kok-Chor. *Cosmopolitanism, nationalism and patriotism*. Cambridge, UK: Cambridge University Press, 2004.

Tarrow, Sidney. *The new transnational activism*. New York: Cambridge University Press, 2005.

Telford, John, and John Cosgrave. *Joint evaluation of the international response to the Indian Ocean tsunami*. London: Tsunami Evaluation Coalition (TEC), 2006.

Tennant, Vicky, Bernie Doyle, and Rouf Mazou. *Safeguarding humanitariarian space: A review of key challenges for UNHCR*. Geneva: UNHCR, 2010.

Terry, Fiona, *Condemned to repeat? The paradox of humanitarian action*. Ithaca, NY: Cornell University Press, 2002.

Törnblom, Kjell Y., and Riël Vermunt. "Towards an integration of distributive justice, procedural justice, and social resource theories." *Social Justice Research* 20.3 (2007): 312–35.

Transparency International. *Preventing corruption in humanitarian operations*. Berlin: Transparency International, 2010. http://www.transparency.org/publications/publications/humanitarian_handbook_feb_2010.

Trudinger, Cathy, and Ian Enting. "Comparison of formalisms for attributing responsibility for climate change: Non-linearities in the Brazilian Proposal approach." *Climatic Change* 68.1 (2005): 67–99.
Tsunami Evaluation Coalition. "Coordination of international humanitarian assistance in tsunami-affected countries." July 2006.
U.S. Agency for International Development. *USAID Office of Food for Peace Haiti Market Analysis*. August 2010.
U.S. Government Accountability Office. "Agency management of contractors responding to Hurricanes Katrina and Rita." (GAO-06-461R) February 2006.
UDHR (Universal Declaration of Human Rights). Approved and proclaimed by the General Assembly of the United Nations on December 10, 1948, as resolution 217 A (III). http://www.ohchr.org/EN/UDHR/Pages/Language.aspx?LangID=eng.
UNDP. *Central Asia human development report*. New York: UNDP, 2005.
UNFCCC. *Kyoto protocol to the United Nations framework convention on climate change adopted at COP3 in Kyoto, Japan, on 11 December 1997*. 1998. http://unfccc.int/resource/docs/convkp/kpeng.pdf.
United Nations. *Millennium project. Investing in development: A practical plan to achieve the millennium development goals*. New York, NY: Earthscan, 2005.
United Nations. *United Nations framework convention on climate change*. London: HMSO Books, 1995: 5.
United States House of Representatives Committee on Government Reform. Minority Staff Special Investigations Division. "Waste, fraud, and abuse in Hurricane Katrina contracts." August 2006: 1–19.
UNOCHA. "Philippines: Typhoon Haiyan action plan—November 2013." http://www.unocha.org/cap/appeals/philippines-typhoon-haiyan-action-plan-november-2013.
Uphoff, Norman, and Anirudh Krishna. "Civil society and public sector institutions: More than a zero-sum relationship." *Public Administration and Development* 24 (2004): 357–72.
U.S. House Government Reform Committee. "Hearings on sifting through Katrina's legal debris: Contracting in the eye of the storm." May 3, 2006.
van de Walle, Nicolas. *African economies and the politics of permanent crisis, 1979–1999*. Cambridge, UK: Cambridge University Press, 2001.
Vasavada, Triparna. "Managing disaster networks in India." *Public Management Review* 15.3 (2013): 363–82.
Villanger, Espen. "Arab foreign aid: Disbursement patterns, aid policies and motives." *Forum for Development Studies* 34.2 (2007): 223–36.
Walster, Elaine, Ellen Berscheid, and William G. Walster. "New directions in equity research." *Journal of Personality and Social Psychology* 25 (1973): 151–76.
Walster, Elaine, Ellen Berscheid, and William G. Walster. *Equity: Theory and research*. Boston: Allyn and Bacon, 1978.

Waltham, Tony. "The flooding of New Orleans." *Geology Today* 21 (2005): 225-31.
Walzer, Michael. *Spheres of justice: A defense of pluralism and equality*. New York: Basic Books, 1983.
Warren, Richard. "The evolution of business legitimacy." *European Business Review* 15.3 (2003): 153-63.
Wartick, Steven L. and Philip L. Cochran. "The evolution of the corporate social performance model." *Academy of Management Review* 10 (1985): 758-69.
Weaver, Gary R., Linda K. Trevino, and Philip L. Cochran. "Corporate ethics programs as control systems: Influences of executive commitment and environmental factors." *Academy of Management Journal* 42.1 (1999): 41-57.
Webster, Donovan. *Fault line: Aid, politics, and blame in post-quake Haiti: Two years after the earthquake, where did the money go?* Retrieved May 16, 2012. Global Post, 2012: http://www.globalpost.com/dispatch/news/regions/americas/haiti/120110/haitiearthquake-aid-rice.
Weick, Karl E. "Reduction of cognitive dissonance through task enhancement and effort expenditure." *Journal of Abnormal and Social Psychology* 69 (1964): 533-39.
Weick, Karl. "The concept of equity in the perception of pay." *Administrative Science Quarterly* 11.3 (1966): 414-39.
White, Richard D., Jr. "Are women more ethical? Recent findings on the effects of gender upon moral development." *Journal of Public Administration Research and Theory* 3 (1999): 459-71.
Wicklund, Robert A., and Jack Williams Brehm. *Perspectives on cognitive dissonance*. Hillsdale, NJ: Lawrence Erlbaum, 1976.
Wiener, Yoash. "The effects of 'task' and 'ego' oriented performance on two kinds of overcompensation inequity." *Organizational Behavior and Human Performance* 5 (1970): 191-208.
Williams, Bernard. "Ethical consistency." In *Proceedings of the Aristotelian society* 39, pp. 103-24, reprinted in Bernard Williams, *Problems of the self.* Cambridge, UK: Cambridge University Press, 1973.
Williams, Bernard. "Persons, character, and morality." In *Identities of person*, edited by Amelie O. Rorty. Berkeley, CA: University of California Press, 1974. Reprinted in Bernard Williams. *Moral luck*. Cambridge: Cambridge University Press, 1981: 1-19.
Winters, Matthew S. "The Obstacles to foreign aid harmonization: Lessons from decentralization support in Indonesia." *St. Comp International Development* 1 (2012): 26. doi:10.1007/s12116-012-911-7.
Woods, Ngaire. "Whose aid? Whose influence? China, emerging donors and the silent revolution in development assistance." *International Affairs* 84.6 (2008): 1205-21.
World Bank. *Global development finance 2007*. Washington, DC: World Bank, 2007.

Yang, Guobin. "A civil society emerges from the earthquake rubble." June 5, 2008. YaleGlobal Online, www.yaleglobal.com.

Zapata, Martí Ricardo. *The 2004 hurricanes in the Caribbean and the tsunami in the Indian Ocean: Lessons and policy changes for development and disaster reduction.* Mexico: United Nations, Economic Commission for Latin America and the Caribbean, 2005.

Zengerle, Patricia. "Will endemic corruption suck away aid to Haiti?" *Reuters*, 26 January 2010. http://www.reuters.com/article/idUSTRE60P3HN20100126.

Index

Active Learning Network for Accountability and Performance in Humanitarian Action (ALNAP), 140–41
Afghanistan, 8, 78, 91
Ajzen, Icek, 31
Anderson, Benedict, 51
Angola, 78, 82
Anti-Corruption Agency (ACA), 98
Aquino, Benigno, III, 35, 103
Aristide, Jean-Bertrand, 71
Aristotle, 39, 40

Bangladesh, 8, 80, 104
Barcan Marcus, Ruth, 27
Barrett, Christopher B., 64–65
basic need approach (BNA), 65–66, 70
best practices, 6, 21, 139–44. *See also* codes of conducts/ethics
Blau, Peter M., 43
Brazil, 81–82, 111
Bueno de Mesquita, Bruce, 14–15
Bush, George W., 92–93

Cahill, Kevin M., 2
Caney, Simon, 155n14
capabilities-oriented approach, 39, 50, 53, 58–59
capacity to pay principle (CPP), 116, 118, 119, 148
CARE USA, 14

Catholic Relief Services (CRS), 13–14, 21–22
Cherpitel, Didier, 63
Chile, earthquake of 2010 in, 4, 8–9, 80
China, 1, 81–82; climate change and, 113, 115
chlorofluorocarbons (CFCs), 106–7
climate-change compensation, 35–37, 103–19, 142–43, 148
Clinton, Bill, 71–72
code(s) of conduct/ethics, 6, 16, 21, 121–44, 149; for climate-change compensation, 139; corporate, 122; and corruption, 138–39; and dependency syndrome, 136–37; of Disaster Response Programmes, 4, 16, 121; and donation fatigue, 137–38; of Humanitarian Accountability Partnership, 4, 16–17, 121, 128–29, 134; of Humanitarian Charter, 4, 121; of IAEM, 20; for institutions' legitimacy, 122–24; of International Red Cross and Red Crescent Movement, 4, 16, 121, 123, 125–27; for leaders, 142, 145–46; of People In Aid, 4, 17, 121, 129, 134
Collier, Paul, 79
Common Emergency and Information System (CECIS), 13, 20–21
Common Humanitarian Action Plans (CHAP), 132

community of justice, 51–53; climate change and, 108–11; corruption and, 95–96; dependency syndrome and, 69–70; donation fatigue and, 82–84
compensation. *See* climate-change compensation
consequence-oriented approach, 4, 7, 39
Coppola, Damon P., 11
corruption, 9, 89–101, 147–48; code of ethics for, 138–39; consequences of, 91; definitions of, 89, 90; donation fatigue and, 34, 82; measures against, 98
Corruption Perception Index, 90, 91
cosmopolitanism, 7, 34; distributive justice and, 46–61; Kant on, 46
Cox, Raymond W., III, 121
Cupit, Geoffrey, 90

Dean, Hartley, 64
dependency syndrome, 7, 63–76, 147; code of ethics for, 136–37; definitions of, 33, 64–66
development assistance, 19, 77–78, 81–82; humanitarian aid versus, 94–95; OECD and, 81–82, 85, 113, 125
Development Assistance Community (DAC), 81–82
difference principle, 57
Disaster Response Programmes, 4, 16, 121
distributive justice. *See* global distributive justice
Doctors Without Borders, 5
donation fatigue, 7–9, 33–34, 77–88; definition of, 34; ethical concerns with, 137–38
Dreher, Axel, 91
Drèze, Jean, 118

Emergency Response Coordination Centre (ERCC), 21
Epictetus, 46
equal consideration of interests, 59
Equal per capita shares (EPCS) principle, 114–17, 119, 148
equity theory, 41–43, 57
ethical-normative approach, 10, 26–37, 90. *See also* codes of conducts/ethics
ethics education, 141
European Union (EU), 20–21, 113–14

Fadlalla, Amal Hassan, 14
Farmer, Paul, 72
Fedderke, Johannes, 91
Federal Emergency Management Agency (FEMA), 92–93
Fineman, Martha, 66
Fishbein, Martin, 31
Fukushima Daiichi nuclear accident, 92, 100

Geneva Conventions, 16, 17, 127
global distributive justice, 39–61, 146–49; evaluation of, 44–46, 48–61; multi-principle approach to, 43–44; preconditions of, 54–56, 111–13; principle of, 56–61; theories of, 10, 39–44
Global Facility for Disaster Reduction and Recovery (GFDRR), 19
Global Resource Dividend (GRD), 58, 94, 98, 99; for climate change, 107, 109–10, 115–16
globalization, 47–48; Pogge on, 59–60; Singer on, 51–52
Good Humanitarian Donorship project, 4, 17, 129–34, 143
greenhouse gases (GHG), 9, 103–6, 111. *See also* climate-change compensation

Index

Guyana, 8

Haiti, earthquake of 2010 in, 4, 8–9, 12, 71–72, 80, 93–94
Hawley, Chris, 93–94
Heilbroner, Robert L., 55
Held, David, 47
Hiroshima bombing, 79
Homans, George C., 42
Howell, Willis, 77
Humanitarian Accountability Partnership (HAP), 4, 16–17, 121, 128–29, 134; Joint Standards Initiative of, 143
Humanitarian Charter, 4, 121, 125–28
Hume, David, 40
hungersite.com, 73
Hurricane Jeanne (2004), 8
Hurricane Katrina (2005), 4, 8, 72–73, 92–93, 104–5
Hurricane Rita (2005), 4, 93
Hyogo Framework for Action (HFA), 18

impartiality, 60, 87, 128; in climate-change compensation, 118; toward corruption, 90–91, 100, 101, 147
India, 1, 81–82, 115
Indian Ocean Tsunami (2004), 4, 8, 67, 79, 98
Indonesia, 1, 82; earthquake of 2004 in, 8, 67, 79
institutions-oriented approach, 39, 57, 74, 88
Integrity Management Committees (IMCs), 98
Intergovernmental Panel on Climate Change (IPCC), 105, 106, 112; "polluter pays" principle of, 114
International Association of Emergency Managers (IAEM), 19–20

International Committee of the Red Cross (ICRC), 5, 131
International Criminal Court (ICC), 117
International Development Association (IDA), 19
International Disaster Response Laws (IDRL), 17–18
International Emergency Management Society, 19, 20
International Federation of Red Cross/Red Crescent Societies, 17–18
international human rights law (IHRL), 22
international humanitarian law (IHL), 15–16, 22
International Monetary Fund (IMF), 12, 47
International Red Cross and Red Crescent Movement, 13, 131, 132; Code of Conduct for, 4, 16, 121, 123, 125–27
Iran, 8
Iraq, 78, 91

Japan, earthquake of 2011 in, 12, 92, 99–100
Joint Standards Initiative (JSI), 143
justice, community of, 51–53, 69–70, 82–84, 95–96, 108–11. *See also* global distributive justice

Kant, Immanuel, 46, 52
Klitgaard, Robert, 91
Kogut, Tehila, 83
Kohlberg, Lawrence, 30
Korea, 81–82
Kurer, Oskar, 90
Kuwait, 81–82
Kyoto Protocol (1997), 105, 106, 114

Lasswell, Harold D., 15

leadership, ethical, 142, 145–46
legitimacy, institutional, 122–24
Lensink, Robert, 64
Lenski, Gerhard, 43
Lentz, Erin, 64–65
Lifton, Robert J., 79

Malaysia, 98
Médecins Sans Frontières, 5
Meyer, Aubrey, 103
Millennium Development Goals (MDGs), 78
multi-principle approach, 41, 43–44
Munck, Ronaldo, 47
mutual assistance principle, 72, 86–87
Myanmar, 12, 91

new humanitarianism, 21–22
Nickerson, Colin, 5
Nigeria, 82
Northouse, Peter G., 145
nuclear power plant accidents, 92, 100

Orbinski, James, 5
Organization for Economic Cooperation and Development (OECD), 81–82, 85, 113, 125
organizational ethics, 121. *See also* codes of conduct/ethics
Oxfam, 12
ozone layer, 107

Pakistan, 8, 80
"People In Aid Code of Good Practice," 4, 17, 121, 129, 134; Audit Toolkit of, 142; Joint Standards Initiative of, 143
"personal heterogeneities," 52–53, 153n3
Philippines, 1, 8, 35; typhoon of 2013 in, 4, 25, 67–68, 103
Plato, 40
Pogge, Thomas, 68, 85, 146–48, 153nn2–4; on charitable giving, 81–82; on climate-change compensation, 107–10, 112, 115–17, 119, 148; on community of justice, 52–53, 69, 84; consequence-oriented approach of, 4, 7; on corruption, 94–95, 97–101, 147; on dependency syndrome, 68–70, 72–75; on donation fatigue, 84–88, 147; global distributive justice theories of, 10, 59–61, 75; on Hurricane Katrina, 73; on "personal heterogeneities," 52–53, 153n3; Rawls and, 49–50, 52, 54, 57–58, 72; rights- and institutions-oriented approach of, 39, 57, 74, 88; on "socioeconomic rights," 50; on veil of ignorance, 54–55; on women's equality, 52–53
"polluter pays" principle (PPP), 111, 113–18, 148
professionalism, 2–3, 149; ethics of, 90, 141, 146; guidelines of, 6, 16, 21; managerialism and, 125

Quality COMPAS, 141

Rawls, John, 49–50, 52, 54, 57–58, 152n1; difference principle of, 72; Sen and, 58; Singer and, 85
Rescher, Nicholas, 43
Rest, James R., 30–31
Riddell, Roger, 65
rights-oriented approach, 39, 57, 74, 88
Ritov, Ilana, 83
Roemer, John, 40
Rothstein, Bo, 90
Rubenstein, Jennifer, 14, 15

Sabbagh, Clara, 44
Sarkozy, Nicolas, 100
Sartre, Jean-Paul, 27–28

Saudi Arabia, 81–82
Save the Children, 83
Schuller, Mark, 93
Sen, Amartya, 68–69, 85, 146–48; capabilities-oriented approach of, 39, 50, 53, 58–59; on climate-change compensation, 108, 110–11, 116–19, 148; on community of justice, 69, 84; consequence-oriented approach of, 4, 7; on corruption, 95–97, 99–101, 147–48; on dependency syndrome, 68–71, 73–75; on donation fatigue, 82, 84, 87, 88, 147; global distributive justice theories of, 10, 56, 58–61, 74, 75; on "open impartiality," 60, 87, 118, 148; on "personal heterogeneities," 153n3; Rawls and, 58
Singer, Peter, 68, 74, 146–48; on charitable giving, 80–81, 87–88; on climate-change compensation, 106–9, 111–12, 114–17, 119, 148; on community of justice, 51–52, 69, 83–84; consequence-oriented approach of, 4, 7, 39; on corruption, 94–98, 101, 147; on dependency syndrome, 68–70, 72, 74, 75, 147; on donation fatigue, 80–81, 83–88, 147; on equal consideration of interests, 59; global distributive justice theories of, 10, 48–49, 56–57, 59–61, 75; on globalization, 51–52; mutual assistance principle of, 86–87; Rawls and, 85
Slim, Hugo, 22, 25
Slovic, Paul, 79
Small, Debora A., 83
Smith, Adam, 55
Sobhan, Rehman, 65
social justice judgments (SJJ), 44
"socioeconomic rights," 50
Somalia, 91

South Africa, 123
Sphere Project, 4, 16, 121, 125–28; Joint Standards Initiative of, 143
Stewart, France, 65
Stiglitz, Joseph, 35
Stoicism, 46
Streeten, Paul, 65–66
Styron, William, 27
Suchman, Mark, 122

Taiwan, earthquake of 2010 in, 8–9
Talmud, 40
Tan, Kok-Chor, 47
Teorell, Jan, 90
Tokyo International Conference on African Development (1993), 77–78
Transparency International (TI), 89, 90, 91

United Arab Emirates, 81–82
United Nations, 131, 132; Angola Verification Mission of, 78; definition of natural disaster by, 1, 13; Framework Convention on Climate Change of, 103, 105, 106; International Strategy for Disaster Reduction of, 18, 19; Office for the Coordination of Humanitarian Affairs (OCHA) of, 17; Office of the Special Advisor on Africa, 77–78
United Nations Environment Programme (UNEP), 105
Universal Declaration on Human Rights (UNDHR), 50, 57
universalizability, 117–18

Venezuela, 81–82

Walzer, Michael, 152n1
Weaver, Gary R., 122
White, Howard, 64
Wingspread Principles, 123

Winters, Matthew S., 15
World Bank, 19, 47; aid to developing countries and, 77–78; anti-corruption mechanism of, 82; disaster aid from, 123
World Meteorological Organization (WMO), 105

Zambia, 79–80
Zengerle, Patricia, 93